The Final E

The Final Energy Crisis

Second Edition

Edited by
Sheila Newman

LONDON • ANN ARBOR, MI

First published 2005. Second edition published 2008 by Pluto Press
345 Archway Road, London N6 5AA
and 839 Greene Street, Ann Arbor, MI 48106

www.plutobooks.com

British Library Cataloguing in Publication Data
A catalogue record for this book is available from the British Library

ISBN 978 0 7453 2718 1 Hardback
ISBN 978 0 7453 2717 4 Paperback

Library of Congress Cataloging in Publication Data applied for

10 9 8 7 6 5 4 3 2 1

Designed and produced for Pluto Press by
Chase Publishing Services Ltd, Fortescue, Sidmouth, EX10 9QG, England
Typeset from disk by Stanford DTP Services, Northampton, England
Printed and bound in the European Union by
CPI Antony Rowe, Chippenham and Eastbourne

Contents

List of Tables and Figures

TABLES

FIGURES

Acknowledgments

Many, many thanks to all the authors in this second edition of *The Final Energy Crisis*, all of whom gave generously of their time and skill and trusted me with a project they feel is of great importance. Antony Boys's knowledge and skills were extra helpful with some difficult early editing and later proof-reading. Special mention should be made of Andrew McKillop, the brains behind the first edition, who found time in a very busy schedule to contribute several chapters to this one. Michael Dittmar would like to thank William Tamblyn – who is also the moderator of EnergyRoundTable (ERT) – for helpful discussion of his chapter. I would like to acknowledge all those cyber-mates in Running on Empty Oz (Roeoz) (moderated by Mike Stasse), and on EnergyResources (moderated by Tom Robinson), among other great energy lists, over the years.

Thanks also to www.energybulletin.net for delivering the news, to www.theoildrum.com for high-quality, in-depth, independent reporting, and to journalist Kellia Ramares, especially for her pioneering work on women and peak oil, as well as her great radio interviews. Jay Hanson's contribution to international debate on energy resources and evolution is acknowledged elsewhere in this volume.

The editor was also helped and supported by many others, but, notably, Jill Quirk, President of Sustainable Population Australia (Victorian Branch); James Sinnamon, Editor of www.candobetter.org, who let me work out some things using interactive pages on his very interesting site; Ilan Goldman, who keeps me up to date with gold, oil, land prices, and permaculture; and John Newman, who did a lot of background reading and introduced me to science and nature. Last, but not least, I would like to thank my canine associates, Nero and Bianca, who made sure I got exercise.

Units of Measurement

1 ton = 2,000 pounds, or 907.18 kg (US short ton)
1 tonne = 1,000 kg, or 2,204 pounds (US metric ton)

1 kilojoule (kJ) = 1,000 or 10^3 joules
1 megajoule (MJ) = 1,000,000 or 10^6 joules
1 gigajoule (GJ) = 1,000,000,000 or 10^9 joules
1 terajoule (TJ) = 1,000,000,000,000 or 10^{12} joules
1 petajoule (PJ) = 1,000,000,000,000,000 or 10^{15} joules
1 exajoule (EJ) = 1,000,000,000,000,000,000 or 10^{18} joules
1 zettajoule (ZJ) = 1,000,000,000,000,000,000,000 or 10^{21} joules

1 watt (W) = 1 joule per second
1 kilowatt (kW) = 1,000 or 10^3 joules per second
1 megawatt (MW) = 1,000,000 or 10^6 joules per second
1 gigawatt (GW) = 1,000,000,000 or 10^9 joules per second
1 terawatt (TW) = 1,000,000,000,000 or 10^{12} joules per second

1
Introduction – In the Beginning

Sheila Newman

In the Beginning ...
The earth's atmosphere was unbreathable to humans. But that was okay, since there were no humans. Photosynthesizing cyanobacteria used sunlight to convert carbon dioxide and water into food, incidentally producing oxygen. The many microbially-mediated rocks (stromatolites) the bacteria left behind from their halcyon days indicate a cyanobacteria population explosion so vast that it seems likely that simple metabolism accidentally transformed the atmosphere to the one we love and overuse today. This "oxygen holocaust" probably also brought about the fossil status of our inadvertent benefactors around 2.5 billion years ago.

Incredibly the evolutionary serendipity from our point of view did not end there because cyanobacteria fats eventually formed the petroleum hydrocarbons which drive the sophisticated combustion engines of trains and boats and planes today.

Yes, if you hadn't already thought about it, the petroleum which fueled twentieth-century cleverness comes out of vast microbial cemeteries in the earth, most of which were formed during two periods of global warming 90 and 150 million years ago. Under pressure from subsequent sedimentation and other geological events, the bacterial corpses cooked, compressed and liquefacted, changing their chemical qualities. Many millions of years later, they became oil. While some of these processes continue today, no new oil reserves could have been created in the short time humans have been using this stuff up.

It is generally estimated that we humans have gone through about half the oil on earth but that our population and our economies have grown so huge that we will use up the most accessible part of the remainder in fewer than 30 years.

Before humans used petroleum for commercial fuel, they used coal. Before they used coal, they used wood. It is believed that the exhaustion of wood supplies led to the use of coal in Britain, where an abundance of iron occurred close to major coal deposits, which provided capital for the industrial revolution. The commercialization of coal accompanied the first big global leap in human population numbers and material consumption, and the formation of corporate structures, from around 1730, until coal was supplemented and overshadowed by petroleum. Petroleum came just in time, since widescale use of coal caused obvious pollution and sickness, visible in pea-soup fogs and in lung and other diseases.

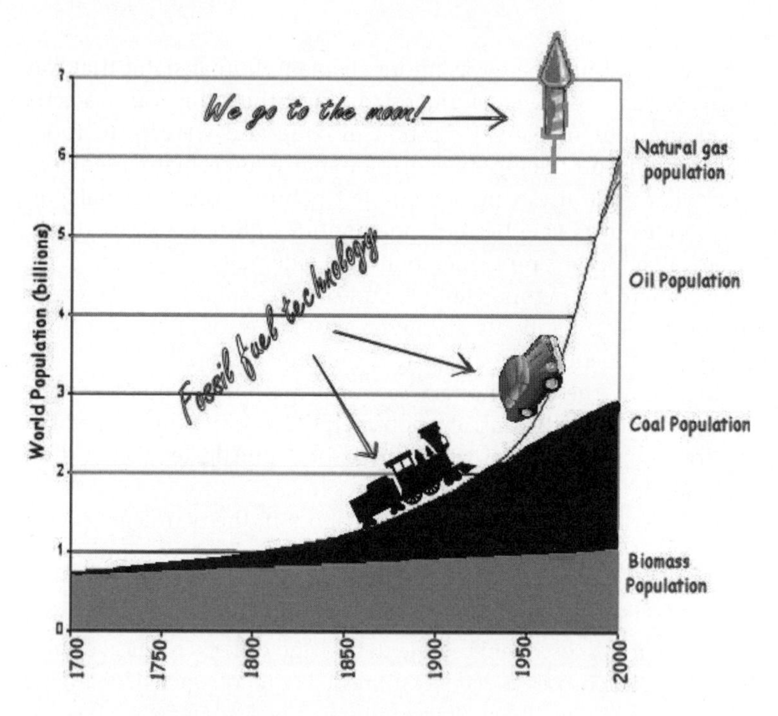

Figure 1.1 Sum-of-energies model of world population

Source: Adapted from Graham Zabel's "Population and Energy," www.dieoff.com/page199.htm.

The association of specific technologies with specific fuels, such as horse power with biomass (vegetable fuel), trains with coal, and cars, planes, and rockets with petrol and gas is fairly obvious. But because we think of our species as independent of natural laws, few of us associate human mass with fuel mass. The concept is as surprising at first as the concept that observed sub-atomic particles, like quarks and neutrinos, react to their human observers. We are not surprised to find that sub-atomic particles react to each other, but we forget that we are ourselves composed of sub-atomic particles. Similarly, we easily accept that abundant fossil fuel enriches society, but it does not occur to most of us that our mass and destiny reflect availability of fossil fuel. That is, that the shape and mass of human population is sculpted by coal and oil.

Around 1998 the concept of a final energy crisis (which had faded right into the background, particularly in the countries which adopted Reagan-Thatcherite economic policies after the 1973 oil shock) started to come back on the Anglophone radar. In my own experience of this, it returned initially on internet groups, led by Hawaii-based Jay Hanson, a man with cyber-charisma even in plain text. He formed the Darwin List around this time and then the Die-off List and then EnergyResources. He should be acknowledged for kicking off an important new energy science-focused intellectual movement on the emerging global internet. In his usual terse but effective style, he summarized the rules and our predicament at www.dieoff.org/synopsis.htm:

> **Energy** is the capacity to do work[1] (no energy = no work). Thus, the global economy is 100 percent dependent on energy – it always has been, and it always will be.
>
> **The first law of thermodynamics** tells us that neither capital nor labor nor technology can "create" energy. Instead, available energy must be spent to transform existing matter (e.g., oil), or to divert an existing energy flow (e.g., wind) into more available energy. There are no exceptions to the thermodynamic laws!
>
> **The second law of thermodynamics** tells us that energy is wasted at every step in the economic process. The engines that actually do the work in our economy (so-called "heat engines"; e.g., diesel engines) waste more than 50 percent of the energy contained in their fuel.
>
> **Energy "resources" must** produce more energy than they consume, otherwise they are called "sinks" (this is known as the "net energy" principle). About 735 joules of energy is required to lift 15 kg of oil 5 meters out of the ground just to overcome gravity – and the higher the lift, the greater the energy requirements. The most concentrated and most accessible oil is produced first; thereafter, more and more energy is required to find and produce oil. At some point, more energy is spent finding and producing oil than the energy recovered – and the "resource" has become a "sink".
>
> There is an enormous difference between the net energy of the "highly-concentrated" fossil fuel that powers modern industrial society, and the

"dilute" alternative energy we will be forced to depend upon as fossil fuel resources become sinks.

No so-called "renewable" energy system has the potential to generate more than a tiny fraction of the power now being generated by fossil fuels!

This second edition of *The Final Energy Crisis*, like the first, explores those limits.

Presented here is a combination of topical, interspersed with technical, political, ecological, and economic articles about energy and society, all examining the idea that the twenty-first century isn't going to be a predictable continuation of the twentieth-century.

The book is divided into four parts: "Measuring Our predicament" (in a variety of ways), "Geopolitics" (is geology destiny?), "The Big Picture" (about "big" solutions), and "After Oil" (about how specific countries may deal with fossil fuel depletion.) Each part has its own introduction to give an idea of its purpose.

NOTE

1. "Work" is a scientific term which could be defined in lay terms as the ability to make something happen by exerting force. "Without energy, nothing would ever change, nothing would ever happen," writes D. Watson, in "FT Exploring Science and Technology," www.ftexploring.com/energy/definition.html.

Part I: Measuring Our Predicament

This part is about oil production peaks, both per capita and total world, and how they may be measured. It relates scientific scales to geography, politics and lifestyle, ideology and philosophy.

"101 Views from Hubbert's Peak" is a socio-political view of what has really been happening since the first oil shock of 1973. Has the world really been doing as well as many economists and politicians would have us believe? And, if not, must we all really head full-speed towards the apocalypse, or might there actually be another route to somewhere better, which has been overlooked because it doesn't suit the current ideology?

In "Prediction of World Peak Oil Production," Professor of Engineering Seppo Korpela elegantly explains the theories and sources used to date oil peaks and depletion curves. This is the nuts and bolts of assessing the state of energy supply. His chapter also tells us the history of this fascinating field which is daily growing in importance. Adepts will appreciate the addition of a learned appendix describing M.K. Hubbert's original methodology.

Oil geologist Colin Campbell's classic work, "The Assessment and Importance of Oil Depletion," describes how oil and human society go back together in time, but how that connection, recently stronger than ever before, is coming to an end. He at once evokes the primordial power of oil and the pitiful economic religion of *growth* with its dogma of *supply and demand*. His background in earth sciences and years of experience in international oil exploration put human demands and expectations into perspective against geological time. This is the voice of experience at the drill-end of real wealth as opposed to hedge funds built on hype.

What is as crucial as oil? Well, coal, if it is to replace much of what we do with oil. At the end of this part, in a chapter written specially for this volume, "Coal Resources of the World," Seppo Korpela evaluates and writes about coal peak estimations for world regions, including the United States and Russia. Most importantly, he publishes for the first time his calculation of the coal peak and depletion curve for China, the world's biggest coal producer and its biggest consumer. Despite massive production, China has no spare capacity. What does this mean for the rest of the world?

2

101 Views from Hubbert's Peak[1]

Sheila Newman

Anyone with a reputation to lose is cautious about pronouncing that global oil production may truly be peaking and that the world may really be heading for petroleum decline. "Peakniks" who sound the alarm every time the price of gasoline goes up risk the same fate as the little boy who cried wolf, with the citizens of the industrialized global community simply rolling their eyes and turning up their MP3 players.

Whilst experienced oil geologists like Jean Laherrère and Colin Campbell assert that global oil reserves estimations are overestimates,[2] industry oil commentators imply that such "pessimism" about oil supply belongs to old men with outdated techniques and ideas. The rejoinder is that they *would* say that, wouldn't they, since they are still working for an industry that needs to push upbeat messages in order to sustain share prices.

The problem is that no one can get hold of a definitive data series on oil production from Mother Earth, and the data we have are commercially sensitive. Since the imagined consequences of oil decline range between a modest readjustment of lifestyle and a rapid descent into pan-cannibalism, catastrophe has to be almost upon us before anyone official is willing to risk causing civil panic, let alone corporate rage and investor alienation.

Nailing the global peak in total oil production is a work in progress, with a declining margin for error, which Seppo Korpela discusses in "Prediction of World Peak Oil Production" in this volume, but what about the *per capita* oil availability peak?

WE ARE DEMONSTRABLY WELL PAST THE PER CAPITA OIL AVAILABILITY PEAK

Although commercially sourced geological statistics are often unreliable and unverifiable in absolute terms, a wide variety of social indicators lends support to statistical records of worldwide and regional declines in per capita oil availability.[3] By per capita oil availability, I mean the amount of oil produced annually divided by the number of people in the world.

Demographic records combined with historical geological records of oil exploration and extraction show that per capita growth in global petroleum energy extracted (defined with or without refined gas-liquids) began to fall shortly after 1978 and that the angle of decline increased from around 1980. Oil extraction rose again in the 1990s but has not kept up with population growth.

Per capita economic growth decline coincided with per capita oil production decline but this is not widely acknowledged. The Anglophone press, particularly, goes on (with one eye shut against the third world) pretending that "everyone" is getting richer.

But, how could this be when available global per capita oil (or oil and gas) appears to have been on a plateau since 1983?

STATISTICS WHICH INDICATE WIDESPREAD DECLINE IN ECONOMIC GROWTH FROM 1974

Not all economic statistics give the false impression that every day everything is getting better and better. Organization for Economic Cooperation and Development (OECD) economist Angus Maddison, for instance, developed a comparative quantitative framework for estimating economic growth across nations and regions going back to 1820.[4] His statistics have an historic basis and use social as well as material indicators. They measure units of "purchasing power" rather than currency exchange rates.[5] Although Maddison expected economic growth to continue through the twenty-first century and did not attribute economic growth to the availability of fossil fuel, his statistics broadly correlate with fossil fuel trends, independently corroborating expectations founded on peak oil/thermodynamic/resource-based thinking.

Maddison divided world per capita economic growth (which he broke down into a variety of regions) into five phases between 1820 and 1992. According to his data and measures, economic growth was the greatest it had ever been from 1950 to 1973, but has been slowing ever since then. This can be seen in the logarithmic graph of world per capita gross domestic product (GDP) 1950–2003 (Figure 2.1).

Those familiar with peak oil theory will recognize that the great growth period of 1950–73 coincides with the massive expansion of the petroleum-based economy, known in France as "*les trente glorieuses*"

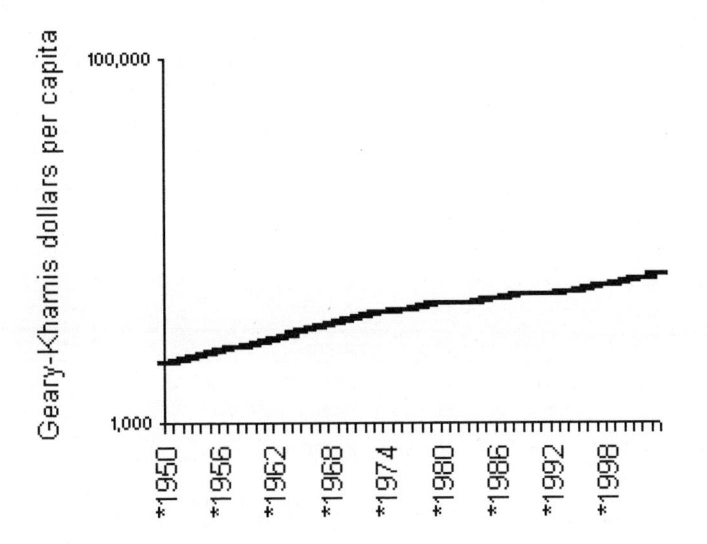

Figure 2.1 World per capita GDP in international Geary-Khamis dollars 1950–2003 (Maddison) (logarithmic scale)

Source: A. Maddison, *Monitoring the World Economy 1820–1992*, OECD, Paris, 2000.

and in English as "the long boom." This period of petroleum-enhanced growth built on another period of unprecedented growth from around 1750, based on that other fossil fuel, coal.

Huge population growth accompanied both these periods. The coincidence of fossil fuel growth and population growth is obvious in Figure 2.2.

The decline of the rate of world per capita economic growth coincided with the first and second oil shocks (approximately 1973 and 1980). Since 1973 per capita economic growth has not "recovered" and unevenness of economic growth has increased.

Despite the conspicuous correlation between growth and fossil fuel resources from the time of the industrial revolution, the dependency of economic growth on fossil fuels is not widely recognized. Angus Maddison himself believed in the economic theory that growth has been "dematerialized" since the first oil shock, based on other statistics which show that calories of energy per production unit have decreased.[6] How could two such opposing evidences coexist in the same world? More about this mystery later.

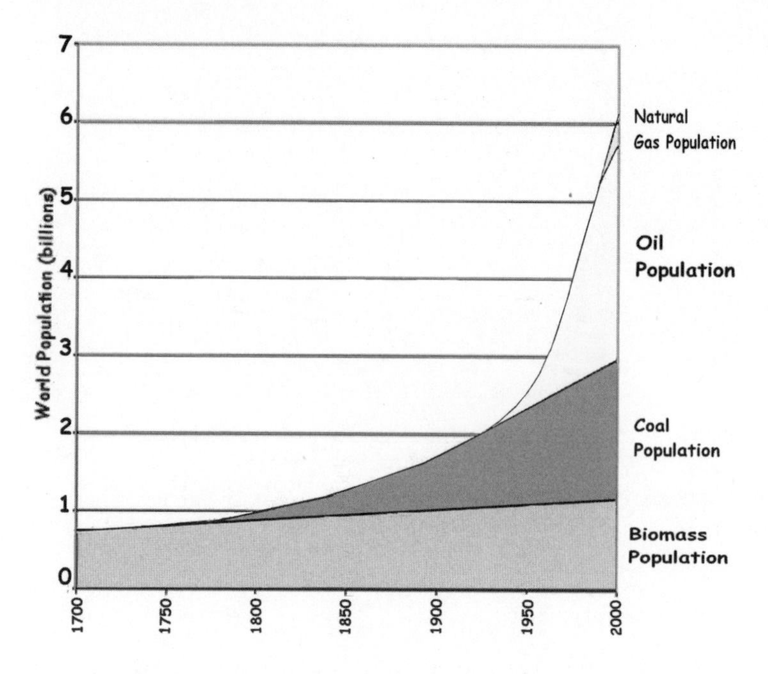

Figure 2.2 Energy sources and world population numbers

Source: Adapted from Graham Zable's "Population and Energy," www.dieoff.com/page199.htm.

POPULATION GROWTH AND OIL PRODUCTION STATISTICS

World population increased between 1979 and 2003 by about 44 percent.[7] Global per capita oil production peaked in 1979 at 0.73 tonnes (US metric tons, or 1,000 kg) per person then dropped by around 14 percent in 1983 to 0.58 tonnes per person, according to British Petroleum's historical timeline series, which starts in 1965.

But, if we were to limit our definition of oil to "crude" (the liquid stuff that comes out of the underground reservoirs plus "condensates" – any associated gases which liquefy under surface atmospheric pressure)[8] which was essentially the pre-1970s definition of oil, the drop would look a lot more severe.

It seems that, in the 1970s, what was generally understood as "oil" was redefined from "crude" to include liquid products of natural gas, as reflected by the two main Anglophone publishers of statistics collected from the industry.

These publishers are the Energy Information Administration (EIA), which publishes the "official energy statistics from the US Government," and British Petroleum (BP). In the EIA statistics the definition of crude is confined to lease condensates[9] and excludes natural gas plant liquids, such as butane and propane, which are recovered at downstream natural gas processing plants or facilities. BP's widely accessible statistics simply appear to include all natural gas as oil. The EIA has another category, "Oil supply," which comprises, as well as crude oil, most other liquids from gas as well as liquids from coal, oil shale, tar sands and bitumen, and includes ethanol.[10]

Where gases have been redefined as oils, there is the semblance of an oil plateau, albeit saddle-shaped, situated between 1988 and 2006 at around 0.6 for BP, and climbing only slightly from 0.65 for EIA "Oil supply," despite its wide definitional catchment. The lower thick black line, which would have been the only measure of our oil supply in the 1960s, has declined overall since 1989. Remember that growing difficulties in oil extraction lower the energy returned on the energy invested (EROEI) for oil, or its net energy value to the world, but increase prices for consumers.

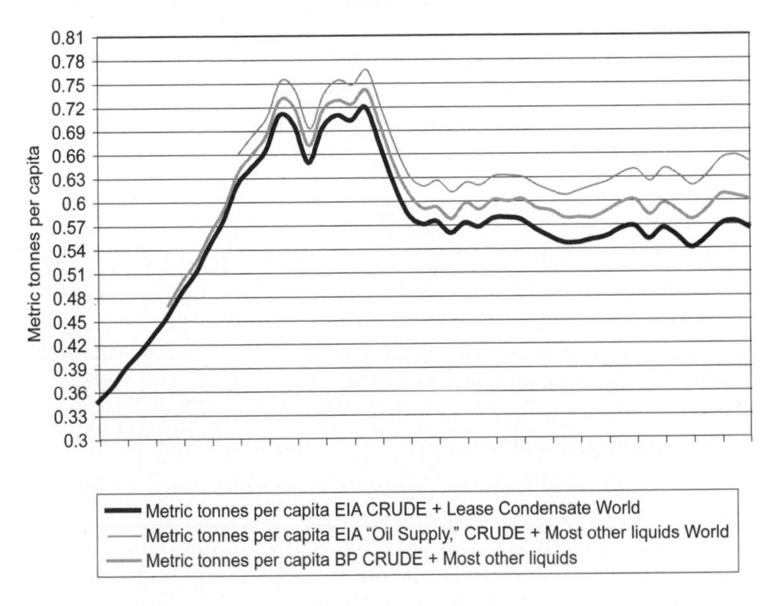

Figure 2.3 Per capita production oil to gas trends 1960–2006

Sources: EIA & BP data.

Figure 2.3 gives three curves, with the lower one (the thick black line) the basic crude (and lease condensates) that the EIA has put out since 1960. The middle (heavy gray) curve represents what BP publishes as "oil," and the highest curve (thin black) shows that the EIA arrives at a somewhat higher figure for gas liquids than BP.

The purpose of this graph is to show how much of the per capita oil plateau has been made up of gas. More every year. (Colin J. Campbell goes more deeply into the production of oil and the production, collection and publication of related statistics in his chapters in this volume.)

For these reasons some oft-quoted oil statistics can make us feel that things aren't as bad as they might seem. It makes sense then to look at other indicators as well – economic, social, and political – which may reflect a reality which the statistics themselves conceal.

NATURAL GAS, ALREADY FILLING IN FOR OIL

Natural gas[11] is a complex of gases, mostly hydrocarbons, which are most often found mixed with, floating on, or near to crude oil reserves. The oil explorers considered it a nuisance. They just wanted to get at the heavy wet stuff that was so convenient to package and sell. So nearly all gases used to be "flared off" (burned off into the atmosphere) because gases are difficult to capture, refine on site, and transport.[12] "Lease condensates" have been counted in with crude for a long time because, although they are gases below the ground, they conveniently condense and liquefy as they emerge into the cooler surface atmosphere.

Now when gas is found with oil reserves it is often or ideally reinjected to increase the pressure at which the oil flows to the well as well as to store it for future use. Since 1970 US industry has become the biggest natural gas user, with households the next biggest. A fact less talked about than oil levels, the US was self-sufficient in gas until the late 1980s, but it now imports more every year.

Despite the greater uptake of natural gas these days, an enormous amount still goes up in smoke without any industrial or domestic benefit to offset its huge contribution to greenhouse gases.

Government, industry, and some environment spokespersons talk nonchalantly about transitioning oil depletion with natural gas before we go on to "alternatives." They tend not to address the fact that this transitioning started in the 1970s. Nor do they mention how quickly the US ran down its supply after it supplemented declining local oil production. No one wanting to reassure voters, customers and shareholders, is likely to speak – even in hushed voices – of the looming gas cliff somewhere in the carbon gas fog not too far up ahead of that luxury liner, *Fossil Fuel.*

DEMATERIALIZATION OR HIGHER QUALITY ENERGY?

There are economic theories to explain how it looks as if some of us got to have our cake and eat it, although they don't talk about cake. They talk about how much more "efficient" we have become. Most rely on the notion that superior technology made it possible to decouple production more and more from fossil fuel resources. We are supposed to be making more with less fuel. But did improvements in technology and better organization really stretch lifestyle and "continuous improvement" seamlessly over the vertiginous plunge in *per capita* petroleum production to deliver its benefits to 1.5 billion more people than in 1979?

Cleveland, Kaufmann, and Stern, in "Aggregation and the Role of Energy in the Economy,"[13] tested the "dematerialization" explanation of conventional economists for claims that economic growth is proceeding well with reduced growth in global oil production.

Looking at the causal relationship between energy use and GDP from 1947 to 1996, they found, instead, that people and business have not used less fuel, but that they have been more careful about the fuels they choose to do different tasks, choosing cheaper fuels to do low production work and more expensive ones for high returns. Cleveland et al.'s main indicator of quality is financial price and it seems to be an indicator which performs well in this case. The authors found that the financial cost of fuels reflects their versatility or adaptability to specific tasks rather than the total calories they embody.[14]

Logistical or engineering factors affect the return on calories. Humans have adapted their social systems and technology to the limitations of fuels and fuel supply. If a fuel must be transported a long way, it makes less sense to use gas, since the cost rises over distance. To keep expensive pipe diameters low, natural gas, for example, needs to be compressed or cooled to liquefied natural gas (LNG) for transport through insulated pipes. (This is also a major reason why using hydrogen as an electricity carrier is so awkward.) You would do better by using a solid (like coal) or a liquid (like petroleum) in this circumstance. Whereas petroleum was once more widely used in North America for domestic heating, because it has a higher calorie content than coal, it now tends to be used for more economically productive work. Coal-fired electricity became a heating option, but now the price of coal is rising due to demand and the cost of transporting it to sprawling populations. Cleveland et al. found that the more expensive the fuel in dollar terms, the more carefully it was used. The authors also found that, even where fuel was cheap to start with, its cost would increase in line with the financial return it provided to the users, so that their financial measure showed reliability over time.

Although the economist Angus Maddison saw that GDP growth was decreasing, he nonetheless anticipated a long period of rapid growth through the twenty-first century, if not so rapid as in the last century. But he based this on his perception of trends of dematerialization of the economy, which he inferred from trends in quantities of calories of oil equivalent used, and failed to account for differences in the nature of fuels and their suitability for different uses. Maddison writes:

> **Past Relation between World Economic Growth and Energy Consumption**
> (3) Tables 4a and 4b compare the growth of world population and GDP with energy use (in terms of both fossil fuels and biomass) from 1820 to 2001. The energy intensity of GDP rose until 1900 (to 0.42 tons of oil equivalent per $1000) and fell in the course of the twentieth century (to 0.27 tons per $1000 in 2001). Per capita energy use at the world level rose about eightfold from 1820 to 2001.[15]

Maddison deduces that technical progress is embodied in machinery (or "technology" advances).[16] Re-examining the situation using Cleveland et al.'s method, one tends to agree with them, that "In economic terms ... technical change has been 'embodied' in the fuels and their associated energy converters."[17]

Maddison's interpretation of the importance of different factors in economic growth supports a deduction that "progress" relies predominantly on human ingenuity. Cleveland et al.'s supports a less gung-ho view. Their view suggests that abundance in quantity and range of fuel types facilitates invention and productivity. A reduction in quantity and variety would send all costs of production up and decrease the opportunities for innovation.

Maddison's optimistic economic perspective is shared by many economists who currently furnish popular and government opinion and this explains perhaps the difficulty that so many people have in considering that the world may be running into some very difficult problems which are likely to overwhelm human ingenuity.

Furthermore, because it suggests (wrongly) that more can be done now with fewer calories of fuel, due to improvements in machinery (rather than due to improvements in selection of fuel type and form), this perspective encourages a view that the world may be steadily diminishing its reliance on aggregate fossil fuel, when in fact, as Maddison himself writes, "per capita energy use at the world level rose about eightfold from 1820 to 2001."[18]

SKEWED DISTRIBUTION MAY DISTORT THE REAL ECONOMIC PICTURE

Another factor which complicates the assessment of material "progress" and its relationship to petroleum fuels is that petroleum fuels are not distributed by the market equally and evenly. Per capita oil is no reflection of actual distribution among different classes of human beings in different countries. Although there has been a world oil market that redistributes oils at the same price to all buyers, not all currency is equal. Ironically, in some places where oil is still an export commodity, such as Black Africa or Latin America, despite the global bull market, much of the population lives in absolute poverty, deprived of land and forced to sell their labor for less than it takes to nourish and house them. This resembles the situation at the height of the British industrial revolution where agriculture, nature, and earlier land tenures and occupations were obliterated by the coal mining and iron works that dominated the lives of those they dispossessed, who had only their labor to sell and whose only prospect of improvement relied on the market for child labor.

So, as the purchasing power is not distributed evenly, neither is the declining per capita wealth. The third world commodity-producing countries assist the richer countries to maintain a certain pace of material acquisition and the currency values to purchase the energy to do so. This suits the corporate world of finance, which keeps the illusion of continuous material progress going by financing national debts on a per capita basis. In those debt-ridden first world countries which have maintained rapid population growth by a variety of political means, including extremely high immigration, the middle classes borrow on their homes or sell off their land to keep up with the rising cost of living, effectively disinheriting their children in exchange for ready cash.

Not all the industrialized countries of the world have continued to grow their populations and to expand their infrastructure. This method of keeping afloat seems to be a specialty of the English-speaking countries like Canada, Australia, New Zealand, and the US. Britain, where their land tenure and inheritance systems arose, continues to pursue a land-speculation economy but may eventually be reined in by its developing dependence on the European Union. With the debt economies go the Thatcher-Reaganite social policies which permit a continuing uneven distribution of wealth within those countries.

There is another way though. Western Europe behaved differently from the Anglophone countries after the first oil shock. It reacted as if this was a practice for the first real oil shortage and may have perceived it as the beginning of a true decline in oil availability.[19] In Germany, Belgium, and France in 1973 the governments drastically reduced immigration overnight, with the cooperation of unions and employers. Spontaneously, the birthrate fell. Postwar plans for major expansion and population growth were abandoned. Public and subsidized private loans for housing and other infrastructure were slashed, causing a dramatic fall in numbers of businesses associated with construction and land sales. Reduction in fuel use, with insulation, renovation and consolidation became core policy.

Why the difference in response between the US and continental Western Europe?

Different factors prevailed. Western continental Europe had experienced fuel shortages during World War I and II (and in wars before those) and again during the Suez Canal crisis. Not being a pioneering culture formed on the expectation of windfalls, Western European culture includes safety valves which US, Australian and New Zealand systems, and even Britain's – all adapted to geological wealth – have failed to incorporate.

Under President Nixon the United States attempted to introduce a formal population policy and energy conservation, as a result of a National Security Study undertaken in 1974,[20] but the corporate machine swung into action there as well.

The Australian Prime Minister of the time, Gough Whitlam, also behaved as if the end of oil was nigh, but the Australian body politic was quickly convinced otherwise by the interests of the privatized property and infrastructure development industry, and the multitude of dependent industries upstream and downstream, including mining, forestry, and finance.

Housing and land development, once major public portfolios in the US and in Australia, had been privatized during the 1960s. The associated industries and corporations had come to rely on rapid population growth, largely through immigration, which they have gradually institutionalized in Australia and the US for their benefit.

A similar private industry dependency on population growth did not exist in Western continental Europe.[21] Even privately built homes relied on publicly subsidized loans, so policies actively promoting population growth could be reversed by government without the political backlash from big business encountered in the Anglophone states.[22]

ORIGINS OF SUBURBIA

The first major peak oil film of the twenty-first century, Barry Zwicker's US film, *The End of Suburbia*,[23] shows how oil underpinned urban expansion from towns into suburbs which have now grown incredibly vast. Implicit in this tale is the population growth which necessarily accompanied that expansion. Although there is a general feeling that the invention of the car had something to do with the postwar baby boom, even demographers usually don't understand how the car permitted post-World War II suburban development, which in turn permitted extra population growth.

To make the logical connection to petroleum rather than fumbling accidental conceptions in drive-ins is difficult for people already thinking wholly within the artificial paradigm of suburbia. The commercially restricted focus of the social sciences from which planners and urban policy makers come, means that only their unconventional members will stray off-limits to make connections in areas like geology and ecology.

Perhaps even then I am failing to make the connection clear. Without the invention of the horseless carriage and plentiful liquid fossil fuel, few people would ever have moved permanently any further than their feet or a horse might take them in a day, which is to say, something like six kilometers. Social opportunities and housing for fruitful unions would have remained limited by distance. Electric rail had already extended the reach of human populations considerably, but dense settlements were still closely oriented to the railway grid.

DIFFERENT SYSTEMS, DIFFERENT OUTCOMES

Anglophone peak oil literature, especially US literature, is filled with the message that the whole world is in the grip of growth ideology. The assumption is that the expansive mode which grips Anglophone states is the same everywhere.

In fact, not every state economy relies on population growth or infrastructure growth. Although the horseless carriage also became popular in Western continental Europe, suburbia never metastasized there to the same degree as in the US because, as I have intimated, housing was seen as a right of citizens which the government had a duty to provide, and which must be budgeted from the public purse. After the two world wars, when towns and portions of cities had been reduced to rubble, housing had lagged behind need. Napoleonic Code inheritance laws tended to keep land within families rather than fragmenting it for sale; there were barriers to the re-zoning of agricultural land, and taxes severely penalized speculative profit on land sales. These and other elements of the prevailing continental European system of land-use planning meant that the new commercial drive for endless expansion, requiring enormous quantities of materials and fuel, driven by population growth, facilitated by the plans of Petainist demographers and economists in de Gaulle's postwar policy committees, fell at the first real hurdle – the 1973 oil shock. Immigration, conception and construction numbers all contracted drastically overnight in Western Europe. Sealing this program were immediate hikes in local taxes on gasoline prices. It was as if Western continental Europeans, protected by a language barrier from direct inoculation with Anglophone hype about progress and the commercialized Protestant work ethic that goes with it, were instantly able to see their geopolitical position, past and future.

This point of view was driven at the highest levels, as this interview with French government minister André Postel-Vinay reveals:

> The doubling of third world population that we are led to expect by the end of the century, gives rise, in my opinion, to considerable danger ... Unless development skills and technology were to improve at an astonishing rate, unless the spirit of human solidarity were to spread in a quite novel way, the proliferation of the human species will increase poverty and malnourishment over vast territories ...
>
> To reject, on principle, the idea of stopping or limiting the entry of new immigrants, would mean that we must allow immigration to increase, even if

> it adds to unemployment and the peopling of slums. This is indefensible ... I am aware of the shocking character of any kind of interruption to or limitation of arrivals, the inhumane quality of sending people back to poverty; but that poverty, alas, is likely to constantly increase and to reach geographical regions ever further away: we cannot take in an unlimited amount. We would perish without relieving it.[24]

On Postel-Vinay's advice, an announcement was circulated immediately by France to immigrant-sending non-European Economic Community (EEC) states warning them of France's decision to stop immigration and of the problems intending immigrants would encounter if they attempted to find work, housing, and visas in France. At the same time policies to consolidate the rights and amenities of immigrants already present in France were initiated.[25]

The French government did not add to the problem by borrowing funds to continue unsustainable growth. It did not adopt Reagan-Thatcherite policies of blaming its marginalized citizens so as to abandon them and ditch its public obligations to house, educate, and provide health care. In US director Michael Moore's 2007 movie, *Sicko*, an expatriate American living in Paris describes the difference as being that, in France, the government is afraid of the people; in the US, the people are afraid of the government.

There are strong international pressures to change the European system, however, to something more favorable to the commercial growth lobby. Recent changes to Napoleonic Law by President Sarcozy look to me like capitulation to this pressure.[26]

Taking advantage of the internet, land speculation has globalized and land-sharks used to the Anglophone system are testing the defenses of every country in the world. Land speculation depends on creating demand, and so land-bankers (people who buy rural land cheaply and organize to re-zone and subdivide it when population and demand have grown in the area) favor population growth above all other things. The corporatized property development system mindlessly pushes against European and other systemic barriers to the market imperative to expand and colonize every part of planet earth for commercial use. Missionaries are hardly necessary when hard cash for land will win over any indigenous landowner as long as he may legally sell his land. So, in those countries where land is still guaranteed to citizens through inherited land tenure, World Bank economists work day and night to privatize village land held safe for centuries in places like New Guinea, Fiji, and even Easter

Island. Land is being touted over the internet from every corner of the globe, from Costa Rica to Cambodia, from South Africa to southern France, from Australia to New Zealand, selling to the highest bidder and creating two economies for every country and island; that of the land-rich and that of the land-poor. In the end the winners are the banks and the people who own them, big and powerful enough to affect national and international laws.

Even the United Nations (UN), once formed to assist postwar refugees, now seems to push policies which could only benefit the corporate world. For instance, the UN suggested in its 2000 press release "Replacement Migration: Is it a Solution to Declining and Ageing Populations?" that Western Europe should replace its population "loss" and maintain a baby-boomer age-to-youth ratio through a massive immigration program from the "developing" world.[27]

This proposal was countered logically and amusingly in an article by French demographer Henri Leridon, who asked, "Who will replace the populations of the third world when they start to age, especially if we have imported most of their youth?"[28] This article also pointed out that to maintain a youthfully structured population indefinitely would require continuous increasing influxes of youthful populations from elsewhere – a completely absurd proposition.

But no mainstream media Anglophone source commented on the absurdity or sought to wonder what was behind this crazy proposal. Who stands to benefit from continuing population growth? Big business.

Big business cannot contemplate an end to expansion and endless expansion depends on population growth continuing forever.

Over the last century, big business has merged seamlessly with mainstream media, using it as a public relations tool to bend social attitudes for commercial ends. Mediatized societies are so dizzy that they can be led to believe that it is reasonable to want to preserve populations in their baby-boom numbers forever simply because the message comes from a media-anointed authority. Industrial man is so isolated from himself and his peers that he needs to believe an impersonal message delivered by a cheerily smiling actor or father-figure politician more than he will believe his own eyes. This is why those societies at the mercy of growthist corporatized government cannot organize to stop population growth, excessive carbon emissions, rapid drawdown on petroleum, privatization of water, privatization of plant and animal genes for agriculture and medicine,[29] massive-scale open-cut coal mining, the razing of local and overseas forests or the destruction of soil. To effectively reduce any of those activities would run counter to the will of the corporate machine. Because all those activities deliver dollars to shareholders and pay for corporate salaries and keep dominant apes dominant in government and other spheres of influence. That we all pay the price so that the corporate machine may profit is not clear to most people and so they do not recognize a common cause and organize against the thing that drives growth because it profits from growth.

"DOOMSAYERS"

Going by global per capita oil statistics, it looks as if the so-called "doomsayers" of the 1970s[30] were right about trends in oil consumption running down supply relative to population on a global basis.

Mathew Simmons writes, in "Revisiting The Limits to Growth,"[31]

> The world is now 30 years into this 100-year view. It did grow as fast as the book warned. The gap between rich and poor never narrowed. Instead, the gap between the "haves" and the "have-nots" grew by a significant measure. It is interesting to contemplate how horrified the book's authors would be today, given the population trends that happened post 1972. The current strain on many of our precious resources is already becoming serious. It would have been far worse by 2000, given the rate of expansion which happened to the world's poor population, had these people also begun to significantly improve their standard of living at the same time. An accidental safety valve for many potentially scarce resources turned out to be the widening of the rich/poor gap.

The "doomsayers" were also right about costs rising with demand in that petroleum would become more costly to a large proportion of peoples whose currency was virtually worthless, and they were right that less oil would be used and that people would suffer because their standard of living would not keep up with population growth. Perhaps they did not actually anticipate that the higher financial cost would lead to a restraint in world economic growth, but that some populations, notably the English-speaking settler states, would continue to purchase oil and to increase the amount they used overall and per capita and to maintain rapid population growth, without apparently paying much more, giving the illusion of persistent rapid economic growth, although less sustained and less rapid. But they would do this by going into debt, by selling off equity, by aggressively commodifying land and citizenship on the global market,[32] and by keeping the rich happy and corporatizing democracy into something called the "client state." Although the US diversified its energy resources, demand for oil still increased.

Countries like France which did not borrow and exchange equity, reduced instead their population growth, diversified energy sources, and increased their technological and design efficiency to reduce fuel needs. Industrial relations and citizenship rights were maintained, especially relative to their erosion in the Anglophone countries.[33]

No industrialized country has managed to replace petroleum for its major transportation fuel, but Western Europe considerably increased km per liter in its automobiles and maintains a lesser energy footprint per capita than the US and Australia.

Between 1973 and 1992 labor productivity grew much faster in Western European countries except Sweden[34] than it did in the US and Australia. Since then, dollar productivity per kilogram of oil equivalent in Western Europe has also far outpaced the expansive Anglophone countries like Australia, Canada, and the US.[35]

A case could be made to encourage the citizens of the world's Anglophone states to exert pressure for European social systems to replace their own overly commercialized ones. Filmmaker Michael Moore suggests this in his bid to get US citizens to look at their corrupt health service industry with new eyes. Local, regional, and national solidarity can help to push back against the abstract machine that has democracy between its teeth and is eating the earth alive.

Antony Boys, in his chapter in this volume on the difficulties severe decline in petroleum supplies would pose for Japan, writes that

> By informing ourselves of potential future problems we may also identify potential new options, such as what kind of society we want to build and how we want to get there. We should aim for the transition to a different kind of lifestyle and society to take place in an organized and orderly fashion in which no one need suffer extreme hardship, gross breaches of basic human rights, or starvation.

COUNTERING PROPAGANDA WITH SIMPLE SOLUTIONS

Propaganda about a "dematerialized economy" makes it hard to establish the reality that material industrial productivity is not actually less reliant on burning fossil fuels than it was in the 1970s, and that drawdown on fossil fuels has in fact been multiplied by the needs of much greater populations. Similarly the obvious still needs to be pointed out that increasing productivity means burning more fuel and outputting more pollution, accelerating petroleum depletion, and adding more greenhouse gases. Not only do we not need all the goods we produce for consumption at home or abroad, we do not need the income they bring, and their acquisition is poor compensation for lives given to industry. Wonderful jobs are few and far between. No one wants to give those up. Some people also derive much of their social life from work, but they would derive similar benefits, and perhaps more status and satisfaction, from other community activities. And plenty of people reach a stage of maturity where childlike obedience to workplace regimes in the cause of producing more and more widgets in different colors, or processing more and more customers a day, with unflinching subservience, challenges every natural instinct.

Instead of those complicated international agreements about percentile reductions in emissions over the years to come, which are hardly enforceable or even measurable, remaining mostly in the control of the corporate emitters, the solution lies much closer at hand, and could ultimately be controlled at grassroots levels by the masses themselves. Relocalization is obviously the best way to develop the solidarity and self-sufficiency to reorganize work.

Political commentator and climate activist Clive Hamilton writes in *Growth Fetish*,[36]

> Reduction in working hours is the core demand for the transition to post-growth society. Overwork not only propels overconsumption but is the cause of severe social dysfunction, with ramifications for physical and psychological health as well as family and community life. The natural solution to this is the redistribution of work, a process that could benefit both the unemployed and the overworked.

He remarks that "Moves to limit overwork ... directly confront the obsession with growth at all costs," and talks about the liberation of workers "from the compulsion to earn more than they need."

Because growth is sustained by a constant "barrage of marketing and advertising" Hamilton wants advertising taxed and removed from the public domain, and television broadcast hours limited so as to "allow people to cultivate their relationships, especially with children."

It should be obvious that the slower we work, the more fuel will remain, the less greenhouse gas will be emitted. If the populations which have ballooned to unimaginable proportions since the 1950s were allowed to return (through natural attrition) to more natural sizes by 2050, and the economy permitted to slow, it would take the heat off the planet and us as well. With so much less effort we could make such a positive difference to the planet and to our personal effectiveness.

NOTES

1. With apologies to Hiroshige's "100 Views of [Mount] Edo."
2. See Colin Campbell's chapter, "The Assessment and Importance of Oil Depletion," in this volume.
3. S.M. Newman, *The Growth Lobby and its Absence: The Relationship between the Property Development and Housing Industries and Immigration Policy in Australia and France*, Swinburne University, Victoria, Australia, 2002, chapter 7, introduction, http://tinyurl.com/yt58yh. Two first world countries which have functioned with growing debt since the first oil shock are Australia and the United States.
4. A. Maddison, *Monitoring the World Economy 1820–1992*, OECD, Paris, 2000.
5. A. Maddison, "Causal Influences on Productivity Performance 1820–1992: A Global Perspective," University of Groningen, Netherlands, *Journal of Productivity Analysis* (November 1997), pp. 325–60.
6. A. Maddison, "Evidence Submitted to Select Committee on Economic Affairs," House of Lords, London, for the Inquiry into "Aspects of the Economics of Climate Change," www.ggdc.net/Maddison/, 20 February 2005, p. 2.
7. UN estimates.
8. EIA Energy Glossary, www.eia.doe.gov/glossary/.
9. www.eia.doe.gov/emeu/aer/txt/ptb1105.html and www.eia.doe.gov/emeu/ispc/appc.html.
10. www.bp.com/statisticalreview (2006 workbook). They also publish natural gas production figures as a separate category.
11. www.neo.ne.gov/statshtml/glossary.htm.
12. www.eia.doe.gov/emeu/aer/eh/natgas.html.
13. C.J. Cleveland, R.K. Kaufmann, and D.I. Stern, "Aggregation and the Role of Energy in the Economy," *Ecological Economics*, vol. 32, issue 2 (2000), pp. 301–17, also at www.bu.edu/cees/people/faculty/cutler/articles/Aggregation_role_of_energy.pdf.
14. Neither was any relationship found between the quality of "transformity" (Odum) and the versatility of fuels for human social needs in industrial societies over the period examined. The authors illustrate this with the example of coal: "Users value coal based on its heat content, sulphur content, cost of transportation and other factors that form the complex set of attributes that determine its usefulness relative to other fuels."

15. Maddison, "Evidence Submitted to the Select Committee on Economic Affairs," p. 2.
16. "The ratio of the latter to GDP rose 16-fold in the UK and USA between 1820 and 2001, and in Japan from 1890 to 2001. This increase was linked to the acceleration of technical progress, much of which had to be embodied in machinery." Ibid., p. 1.
17. Cleveland et al., "Aggregation and the Role of Energy in the Economy," p. 313; Douglas Reynolds, "Energy Grades and Historic Economic Growth," www.theoildrum.com/node/2913#more.
18. Maddison, "Evidence Submitted to the Select Committee on Economic Affairs," p. 2.
19. Newman, *The Growth Lobby and its Absence*, chapter 7.
20. Stephen D. Mumford, *The Life and Death of NSSM 200: How the Destruction of Political Will Doomed a U.S. Population Policy*, Center for Research on Population and Security, Research Triangle Park, North Carolina, 1996. See also appendix 2, recommending population growth reduction as a method of combating fossil fuel depletion. NSSM stands for National Security Study Memorandum. This was an interagency study of world population growth, US population growth, and the potential impacts on national security.

    Aristide R. Zolberg, "Are the Industrial Countries Under Siege," in G. Luciani, ed., *Migration Policies in Europe and the United States*, Kluwer Academic Publishers, Netherlands, 1993, pp. 53–82, p. 61: "In the 1970s, mounting objections by conservative segments of the citizenry to the presence of culturally and often somatically distinct minorities, as well as the oil crisis and ensuing economic crisis, prompted the governments of the industrial countries to undertake a drastic reevaluation of ongoing immigration, but the difficulty of reducing the flows to the desired level, as well as to restoring the *status quo*, precipitated renewed fear of 'invasion'. In the United States, in the 'stagflation' 1970s, estimates of illegal immigrants escalated to as high as twenty million, on the basis of which it was argued that the nation had 'lost control of its borders'. The major solution proposed was to impose sanctions on employers of unauthorized labour, but this failed of enactment because of resistance by organized business interests, so that in 1979 the Congress established a commission to overhaul the entire immigration system."
21. Unfortunately the many changes that Nicolas Sarkozy has recently made to the Napoleonic Code, especially regarding inheritance laws, will almost certainly see this change for the worse. See more on this in "France and Australia After Oil," this volume.
22. Newman, *The Growth Lobby and its Absence*, chapters 3 and 7.
23. Barry Zwicker, *The End of Suburbia*, US, 2002, http://www.endofsuburbia.com/.
24. Translation by S.M. Newman. Postel-Vinay gave me a copy of this interview himself, commenting that he had just re-read it and wanted me to know that it was an extremely accurate journalistic account of his reasons for ceasing immigration. He told me this on the occasion of a visit to his Paris apartment to collect copies of documents on the morning of Saturday 13 January 2001. The quotation is of André Postel-Vinay,

Minister responsible for Immigration from May 1974, in an interview with Jean Benoit in *Le Monde*, September 24, 1974, p. 19. As well as being Minister responsible for Immigration at the time, Postel-Vinay had also previously been the main person responsible for administering foreign aid funds for 27 years and was well aware of the burgeoning populations and economic and political problems in the third world. As well as this familiarity with external events, from 1963 to 1974 he had overseen the administration of an organization for immigrant housing in France. In addition, Postel-Vinay also wrote to the executive government committee he advised: "Whilst our housing program has never exceeded 117,000 dwellings in total, for the entire population, French or foreign, the number of families of foreign workers arriving each year in France has risen, on average, to 38,000 since 1970. In a country like ours, where there are a considerable number of slums, the comparison between the two numbers – 117,000 and 38,000 – gives an idea of the tensions and social oppositions that can arise in response to such a high rate of family reunion." (Unpublished proposal to the Comité Restreint of the French government by the Minister responsible for Immigration policy, Postel-Vinay. Copy given to me by the author in Paris on Saturday, January 13, 2001. Hard copy in appendix 6 of Newman, *The Growth Lobby and its Absence*.

25. P. Weil, *La France et ses étrangers, l'aventure d'une politique de l'immigration 1938–1991*, Calmann-Levy, Paris, p. 86.
26. See "France and Australia After Oil," this volume.
27. United Nations Population Division, "Replacement Migration: Is it a Solution to Declining and Ageing Populations?" Press Release, January 6, 2000. Final report released on March 22.
28. Henri Leridon, "Vieillissement démographique et migrations: quand les Nations Unies veulent remplir le tonneau des Danaïdes," *Population et sociétés*, INED, Paris.
29. For a science-based thriller packed with ideas about the relationship between gene privatization and corporate politics, see Michael Crichton, *Next*, HarperCollins, US, 2006.
30. Such as in the Club of Rome's first publication by Donella H. Meadows, Dennis L. Meadows, Jørgen Randers, and William W. Behrens III, *The Limits to Growth*, New York, Universe Books, 1972. Peak oil analyst Matt Simmons writes: "While many readers concocted various 'imaginary' assumptions, the book's conclusions were quite simple. The first conclusion was a view that if present growth trends continued unchanged, a limit to the growth that our planet has enjoyed would be reached sometime within the next 100 years. This would then result in a sudden and uncontrollable decline in both population and industrial capacity" (M.R. Simmons, "Revisiting The Limits to Growth: Could The Club of Rome Have Been Correct, After All?" www.greatchange.org/ov-simmons,club_of_rome_revisted.html).
31. Simmons, "Revisiting The Limits to Growth."
32. S.M. Newman, "Land and Housing Prices and Land-Use Planning in Australia and Elsewhere and the Impact of Globalisation, the Internet, Trends in Natural Increase, Households and Immigration." Submission

to the Productivity Commission's Inquiry on First Home Ownership, www.pc.gov.au/inquiry/housing/subs/sub153.pdf.

33. Ibid.
34. Maddison, *Monitoring the World Economy 1820–1992*, p. 79.
35. "Units: Tonnes of oil equivalent (toe) per million constant 2000 international $," "Energy and Resources – Energy Intensity: Energy consumption per GDP," http://earthtrends.wri.org//searchable_db/static/668-6.csv, accessed August 11, 2007. Original source was World Resources Institute. See also Newman, *The Growth Lobby and its Absence*, appendix 3, http://adt.lib.swin.edu.au/uploads/approved/adt-VSWT20060710.144805/public/06appendix3.pdf.
36. Clive Hamilton, *Growth Fetish*, Allen & Unwin, Australia, 2003; Pluto Press, London, 2005, last chapter, especially, pp. 218–20.

3

Prediction of World Peak Oil Production

Seppo A. Korpela

INTRODUCTION

Those who read articles on oil in the daily papers often encounter the statement: *with present rate consumption oil will last 40 years*. This number is obtained by dividing the reported reserves by annual production. Since the reported reserves are in round numbers, 1,200 giga-barrels (Gb) (thousand million barrels) and 30 Gb is used for annual consumption, the outcome is 40 years.

A 40-year reserve life exceeds the lifespan – and therefore the concern – of most readers. Only an alert reader will see that oil production will not stay flat for 40 years and then suddenly drop to zero. Rather, it will rise to a peak, after which humankind faces an era of declining production. Thus it is the *peak production rate* that is the most important event regarding our future reliance on petroleum. The media would do us all a service by reporting how close this might be.

The subject of predicting peak oil production occupied the mind of M. King Hubbert during most of his professional life. After obtaining an MSc degree in geology from the University of Chicago he moved to a position of a geology instructor at Columbia University in 1930. While there he submitted his PhD dissertation to Chicago in 1937, then moved to Shell Oil Company in 1943, and later became its Associate Director of Exploration and Production Research. Upon his retirement from Shell he joined the United States Geological Survey in 1963 and remained there until 1976. Hubbert's interest in mineral and energy resources began during his college days in 1926. In 1949 he published a paper in the journal *Science* in which he showed concern regarding the finiteness of the petroleum era and among other thoughts advanced a conjecture that the United States is likely to have 10 percent of the world's ultimate endowment of oil. Seven years later, while at Shell Research Laboratory, he delivered a paper, in March of 1956, at the regional meeting of the American Petroleum Institute in San Antonio, Texas. There he predicted that oil production in the United States would peak in about 1966 if the lower of his estimates of 150 Gb of total oil were accurate; or 1971 if the higher estimate of 200 Gb held. These two estimates for *ultimate production* were judged by oil geologists to be reasonable at that time. Oil production, in fact, peaked in the United States in 1970, validating Hubbert's analysis.[1]

The decade of the 1970s was a watershed in the history of energy in the world, as the general public experienced its *first oil shock* in 1973, caused by an Arab oil embargo to those countries that had supported Israel in the Yom Kippur War. The *second oil shock* took place when the Shah of Iran fell from power in 1979. Oil prices reached record heights as the decade of the 1980s began, but opening up of the North Sea and Gulf of Mexico kept production up and prices dropped, even while the Iran–Iraq War raged through most of the 1980s.

In 1973, when the world's ultimate oil endowment was estimated at about 2,000 Gb, only 300 Gb had been used. According to those numbers, peak production would not take place for another 30 years at least. The United Nations study published in 1972 put peak production at about the year 2000.[2] Similarly in 1956, Eugene Ayres thought that peak might occur perhaps in 1970 for United States oil production, 1975 for (natural) gas, 2000 for world petroleum, and 2050 for United States coal.[3]

Astute professionals in the oil industry still placed the world's oil production peak in the future, but noted that new oil discoveries were becoming increasingly difficult. Factoring this into their models, they thought that by the end of the first decade of the twenty-first century oil extraction would begin its decline. After Hubbert's pioneering work, prediction of world future oil production was taken up by a group of oil geologists, the most notable of whom are Colin J. Campbell, Jean H. Laherrère, L.F. Ivanhoe, Walter Youngquist, A.M.S. Bakhtiari, and Kenneth S. Deffeyes.

In the March 1998 issue of the *Scientific American*, C.J. Campbell and J.H. Laherrère published a joint article, "The End of Cheap Oil,"[4] in which they masterfully laid out the foundation for the reading public to understand the situation regarding world oil. Campbell's monograph,[5] which appeared the same year, is a longer exposition of the subject, with exhaustive graphs and tables to support his warning that the era of cheap oil is nearly gone. He alerted his professional colleagues of the problem in the leading industry publication, the *Oil & Gas Journal*.[6]

Campbell divides oil-producing countries into three groups: those that are past their peak production; those that are near the peak but have not quite reached it; and the so-called "swing producers," all located in the Persian Gulf region. Swing producers are called upon to supply that shortfall as other producing countries, one by one, will pass their production peaks and – as their production falls and domestic consumption grows – cease to be oil exporters. Having access to an industry database, Campbell has been able to track production and depletion patterns in the various large petroleum basins of the world. Using this data, he makes yearly assessments of major producers, and reports the results via the Newsletter of the Association for the Study of Peak Oil (ASPO).[7]

Jean Laherrère refined and extended Hubbert's methods. As a consultant to the oil industry, he has access to the same industry database as Campbell does. This enables him to accurately track discovery trends and use them in his analysis. His work locates the world peak for oil *discovery* in the early 1960s, with approximately 300 Gb still to be found, which *at current production rates* would provide about ten years' supply. This 300 Gb represents a mere 15 percent of the world's initial or ultimate oil endowment, meaning that *85 percent has already been discovered.* Laherrère's graphs, which show how cumulative production lags accurately backdated oil discovery by some 38 years, ought to convince anyone of the nearness of the coming oil production peak.[8]

The late L.F. Ivanhoe began estimating the oil and gas potential for the world during the late 1960s. In 1993 he and G.G. Leckie concluded that the 1,131 largest oilfields in the world possessed 94 percent of world's oil. Since their tally included 41,164 oilfields, the importance of the largest fields on the world's supply is plain.[9] In his book *Geodestinies,* Walter Youngquist provides a wealth of information on all aspects of mineral resources. His view today is that the world will never extract more than 90 million barrels of petroleum liquids a day. As a former Exxon geologist, he is also well aware of the slowness of the process of extracting heavy oil from Venezuela and upgrading the bitumen from Alberta in Canada. In fact, he has offered the opinion that by 2030, extracting oil from the oil sands from Canada's vast resource will suffice Canada's needs only.[10] Finally, A.M.S. Bakhtiari, retired senior analyst at the National Iranian Oil Company, who contributes articles to industry publications such as the *Oil & Gas Journal* on reserves in the Middle East,[11] is skeptical that Persian Gulf countries can increase production at the rate suggested by the International Energy Agency, which would demand an increase of net oil exports from the Middle East

OPEC producers from about 19 million barrels/day (Mbd) in 1997, to 46.7 Mbd in 2020.

Deffeyes gives an excellent account of Hubbert's methods for predicting the world oil production curve in his book *Hubbert's Peak*, which is not only a primer in petroleum geology, but a biographical reference on the life of his colleague Hubbert at the Shell Research Laboratory. Deffeyes's book and access to historical data on the internet have prompted many to repeat Hubbert's calculations for the world oil peak, now with the latest data and with the aid of a computer.[12]

In the appendix at the end of this chapter are the mathematical details on how Hubbert carried out his analysis by two different methods. As new data become available each year, constants in each model can be re-estimated, and the accuracy of the prediction thereby improved. For this reason, the models, when applied to the oil production history and outlook for the US, which is in its late phase of decline, are in excellent agreement. With the benefit of US experience, the models can now be used to show that the world's oil production is near its peak today, and that production of all liquid hydrocarbons will peak by 2011, assuming that demand remains flat or increases from now on. These models also predict that the decline rate will progressively increase to a value slightly under 5 percent a year in 30 years. By then the world will be in chaos, and political events, which the models obviously cannot predict, will undoubtedly influence the decline.

In discussing oil production, what to count as oil needs to be clarified. In addition to the oil extracted from oilfields, in natural gas production the heavier components are in liquid form at atmospheric pressure. This is called lease condensate and, as it is liquid, it is often counted as oil. Additional liquids are obtained in natural gas processing plants. Similarly, refinery gains arising from separation of the hydrocarbons of differing molecular weights add to the volume of liquids produced. These are the main reasons that the world oil production today is reported as being either 73 million barrels/day, or the 17 percent higher figure of 85 million barrels/day. These can be converted to tonnes by dividing them by 7.33, for this is the approximate number of barrels in a tonne owing to the variability of density with the grade of crude. Heavy oil, tar-sands, deepwater, and polar oil are more expensive to produce, and although partly in the stream today, are likely to be more important when world conventional oil is in steep decline.

In the recent past, production statistics in the *Oil & Gas Journal* have not shown any significant increase in oil production, reporting a mere 125,000 barrels increase in 2006. The US Energy Information Administration (EIA) which gives statistics for all petroleum liquids, showed an increase of 19,000 barrels a day and the data sheets show that production has so far exceeded 85 million barrels/day only during three months; in May 2005 and in July and August of 2006. The International Energy Agency (IEA), on the other hand, reported petroleum liquids production as having increased by *500,000 barrels/ day* during 2006, revealing how much the measured quantities of oil flows deviate in the tallies of official bodies.[13]

PRODUCTION CURVES FOR THE US

Data to determine US oil production are available from the EIA website[14] and from Campbell's monograph.[15] Hubbert's analysis is given in the appendix at the end of this chapter. It is based on logistic equation and on yearly and cumulative production.

The actual data are plotted in Figure 3.1 which shows that peak production took place in 1970. The subsequent decline to a bottom in 1975 was the result of conservation efforts after the first oil crisis. The rising trend to year 1985 was caused by a drilling boom in the Gulf of Mexico and completion of the Alaska pipeline, both caused by the 405 percent nominal price rise for crude oil between 1973 and 1981. Owing to Alaskan oil, the year of peak production, according to the model, has moved to 1977 and the actual secondary peak to 1985. From a historical perspective this seven-year shift is an exceedingly short interval in time.

The deepwater oil in the Gulf of Mexico may slow the rate of US depletion for the next few years, but setbacks are possible. For instance, Hurricane Katrina stopped oil production from large platforms such as Mars, which was severely damaged, and the start of Exxon's Thunderhorse field was delayed by three years. Once this production starts to diminish the yearly decline rate is likely again to accelerate to the 4.6 percent value predicted by the model. Data show that over the last dozen years the decline rate in the US has been 2 percent a year.

Cumulative oil production is shown in Figure 3.2. The curve tends toward 240 Gb, which is the estimate for the ultimate production. The cumulative amount produced in the US to the end of 2006 is 194 Gb. Since the published reserves are 22 Gb, this leaves 24 Gb to

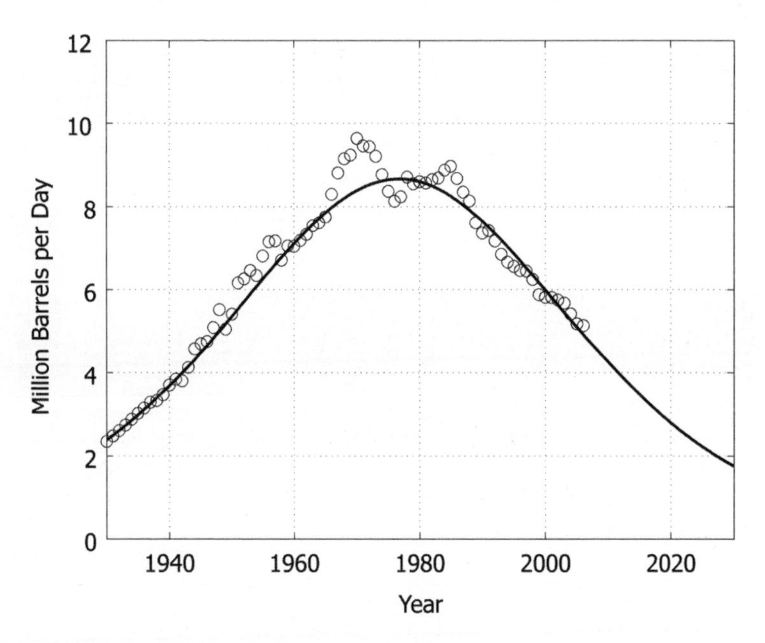

Figure 3.1 Annual oil production in the US

The open circles are the actual data and the solid line represents a calculation by the logistic model.

Sources: EIA data; C.J. Campbell, *The Coming Oil Crisis*, Multi-Science Publishing Company and Petroconsultants S.A., 1998.

a category of *reserve growth* in existing fields and yet-to-find fields.[16] Claiming 95 percent certainty, the United States Geological Survey (USGS) estimates that US ultimate production is 244 Gb, and with only 50 percent certainty, it offers a higher estimate of 362 Gb.[17] The higher estimate is quoted in the newspapers, unfortunately, although the lower number is consistent with the analysis based on logistic equation. The USGS reserve estimates have been controversial in the past and in the 1960s Hubbert himself was faced with ever-increasing estimates for ultimate production, reaching as high as 600 Gb.[18] Hubbert was a member of USGS staff from 1964 to 1976. Reading his 1982 report, one surmises how the ever-increasing reserve estimates must have given him impetus to put his method for estimating the reserve base on a very strong and scientific footing. Even if the latest estimate of USGS has declined substantially, one is at a loss to find a scientific explanation for the high value of 362 Gb still offered to the public.

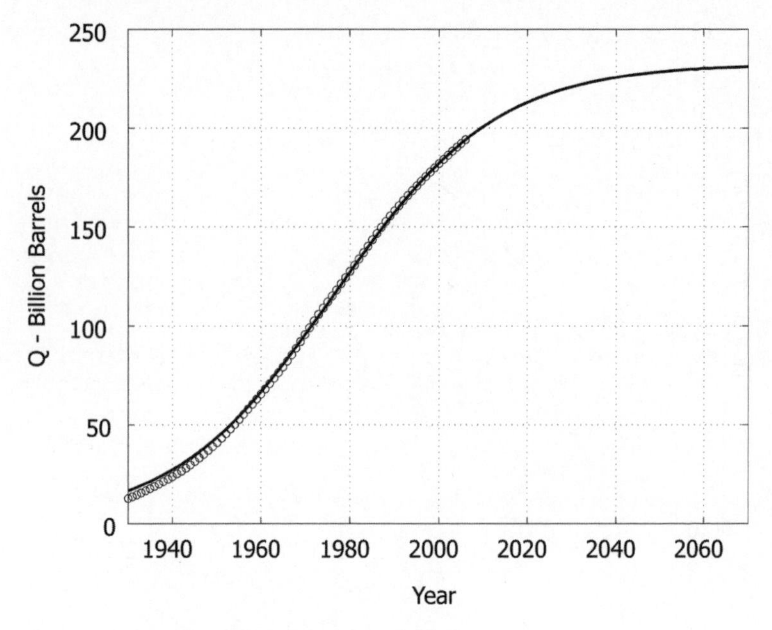

Figure 3.2 Cumulative oil production in the US (Q)

The solid line is the theoretical prediction and open circles are the actual production data. See appendix for detailed explanation.

Source: EIA data.

It is remarkable how well the general trend of the theoretical graph follows the actual data. One might object and say that the excellent fit is primarily owing to data from a long historical record being used to predict the parameters in the model, and it may not have had this predictive capability before this data were available. But, had the data from 1958–66 been used in 1967 to predict the peak, the result would have shown 1976 to have been the peak year for US production, a date six years later than the actual peak.

The key to Hubbert's prediction was his recognition that oil production *must* follow its discovery pattern. Through his study of US discoveries, which peaked during the early 1930s, he was able by 1956 to make his bold predictions concerning US oil; this before he had developed the mathematical basis on which to apply rigor to his forecasting. Although his prediction was dismissed, it has turned out to be accurate. Many today will also dismiss the prediction of world oil peak being imminent, despite the improvement in reserve reporting, analysis, and forecasting techniques, and the fact that the world peak of production is nearly at hand.

WORLD OIL PEAK

The arrival of world oil peak is a much more serious issue than the US peak nearly 40 years ago, for it will usher in the final energy crisis. This section employs the same method as was used for predicting US oil production decay to the question of predicting the date, and volume of world peak oil production.

To obtain the intrinsic growth or decay rate and the ultimate figure for world production, data are plotted in Figure 3.3 in the manner discussed in the appendix. It also contains the US data for purposes of comparison. The scale along the top refers to US production. The trend lines are remarkably similar, validating Hubbert's and Ayres's hunches that the US is likely to have 10 percent of the world's oil.

The calculation uses *Oil & Gas Journal* data and is for oil only. A least squares line drawn through the last 20 years crosses the vertical

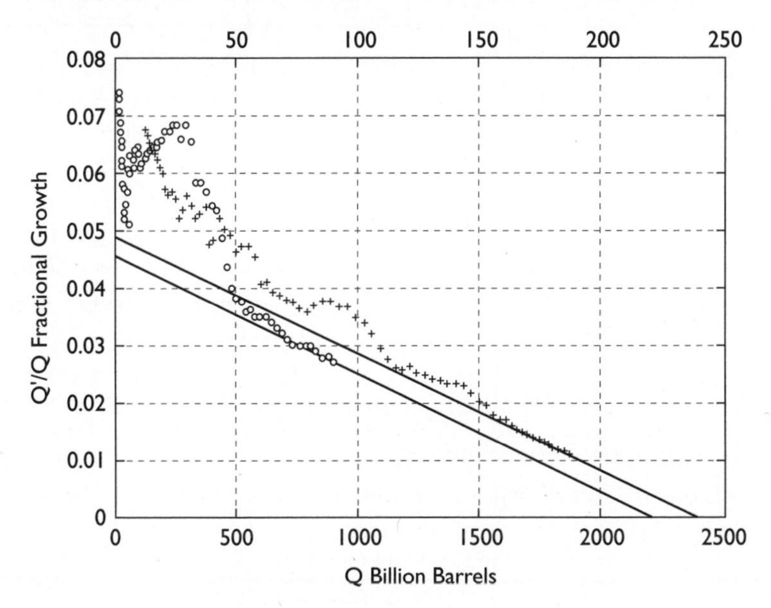

Figure 3.3 A plot to estimate parameters for world oil production

Open circles are for world oil; crosses refer to US oil.

Sources: US data from EIA-DOE; data prior to 1970 for the world from C.J. Campbell, *The Coming Oil Crisis*, Multi-Science Publishing Company and Petroconsultants S.A., 1998.

axis at *0.0472*. This is the intrinsic decay rate and its significance is that, according to the model, after peak oil, the production decline rate will move progressively toward 4.72 percent a year. Technological improvements in oilfield technology may slow the decline somewhat, but not sufficiently to avoid the consequences of severe shortages.

In the calculations in Figure 3.3, ultimate production is situated where the straight line crosses the horizontal axis and today this intersection is at 2,200 billion barrels. 2005 is, so far, the year of peak production. A *2,400 Gb* for the ultimate would delay the theoretical peak until year 2011. However, at the peak, production is quite flat, which means that the demand-driven projection of EIA, requiring a theoretical increase of about 23 percent production increase from 2007 to 2015, is unlikely to be met.[19]

The *2,400 Gb* estimate is about the most generous that the data will permit. It would require many more deepwater discoveries to be made and brought on stream, and at record speed. There was hope that the US intervention in Iraq could supply the shortfall by starting off production from her undeveloped but discovered fields. Given that consumption during 2005 was 27 billion barrels, each 108 Gb addition – which is about the size of Iraqi reserves – to the world's ultimate reserve *delays* the production peak by only about two years, and the ultimate reserve lifetime by four years. Thus even the most prospective country of the world will not appreciably delay the inevitable decline.

A more realistic estimate for the world's ultimate endowment, including yet-to-be-discovered reserves, is 2,200 Gb, as this calculation has shown. This places the theoretical peak production to mid-year 2008. Annual world oil demand growth has been close to 1.8 percent since the early 1990s, and the New Industrial Countries of East and Southeast Asia have typically shown national growth rates for oil consumption of 3.5 to 5 percent per year, and China's growth of oil consumption averaging 7 percent over the last ten years.

In Figure 3.4, annual world oil production plus data for all petroleum liquids production are shown. Whereas total liquids production grew 7 million barrels per day over 2003 and 2004, during the last two years production has been flat. The year 2011 is projected to be the year of peak production for all petroleum liquids. The actual data are already above the theoretical curve, which means that oil production may drop at any time, as it already has on a monthly basis through 2007.

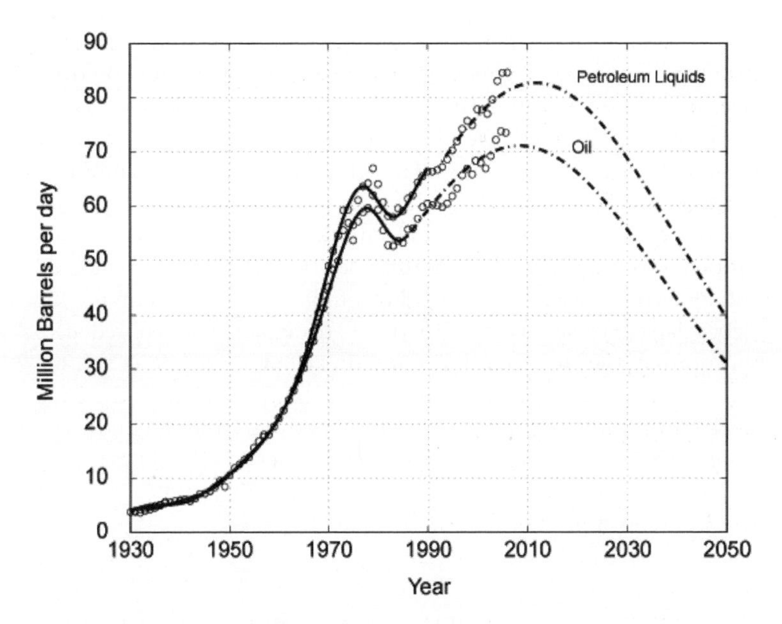

Figure 3.4 Projection for annual world oil production

Theoretical peak production is predicted to take place in 2008, and for all petroleum liquids production in 2011.

Sources: US data from EIA-DOE; data prior to 1970 for the world from C.J. Campbell, *The Coming Oil Crisis*, Multi-Science Publishing Company and Petroconsultants S.A., 1998.

The curve through the data is drawn to aid the eye, using actual data; the curve elements for the last few years and future production are from the logistic model. In fact, only the annual production for the past 17 years, and cumulative production calculated at the start date of 1990, are needed to project future production. This ought to put to rest the objection that the logistic model is "unsuited" to such forecasting because it gives a symmetrical production history.[20] This objection is voiced repeatedly. However, the shape of the past production history is irrelevant, and it does not matter how it fluctuated. The fall and subsequent rise during the 1980s are likewise irrelevant for determining future production. What the model aims to do is to calculate future production from the past production history. When the calculation is carried out each year, a new estimate is made of the ultimate production, as well as the intrinsic decay rate. They show that the year of projected peak production has shifted slightly to the future, but the gap has now essentially closed.[21]

In his yearly assessment of world oil, Campbell situates us on a production plateau for conventional oil, with rising production from deepwater reserves until 2010. Since non-conventional oil, including not only that obtained from deeper parts of the world's sedimentary basins, but also that from the polar regions, is more costly to produce, prices and production during the plateau will be determined by an interplay of political events and economic factors, matters that are amply expounded elsewhere in this book. In fact, by the end of 2007 it had become clear that the world would face a US housing-bubble debt-related recession. Should this happen, demand could be reduced, with oil reserves used up during the next few years at a lesser rate but over a longer period, pre-empting any higher peak than the 2005 peak. The decline of the US dollar has already contributed to the rise in oil price to over US$100 per barrel, and under any hypothesis there is little prospect of cheap and abundant oil remaining a fixture of the world economic situation.

Cumulative production for the world is shown in Figure 3.5. The deviation in the production history for the early periods shown in this graph is a result of estimating the intrinsic decay factor from the last 19 years and ignoring the early period. The left-most curve is an approximate reproduction of the data from Laherrère's paper on cumulative discovery.[22] He obtains it by backdating any reported reserve additions to existing fields, so that they represent oil in the original find, underestimated at first, but revised using new knowledge of the reservoir, as the production progresses and the field size is better delineated. To carry this out, each revision must be assigned to the proper field, which of course means that field-by-field data must be available. Since these data are difficult to obtain, the reader is referred to the original article by Laherrère to see the same figure and his discussion. The discovery history can also be modeled by Hubbert's methods, and since its pattern is similar and precedes that of cumulative production by roughly 40 years, using either the US or world data, how the world's oil production history will evolve is unassailably easy and certain to predict.

This graph and the cumulative production model show the strength of Hubbert's methods for predicting future oil production. The objection that this model is not solidly based on geological fact and, hence not reliable, is misplaced. The claim that the model is too simple to represent the multitude of factors that must drive oil consumption is also without merit. Whereas it is true that the model involves only two parameters, both are estimated each year

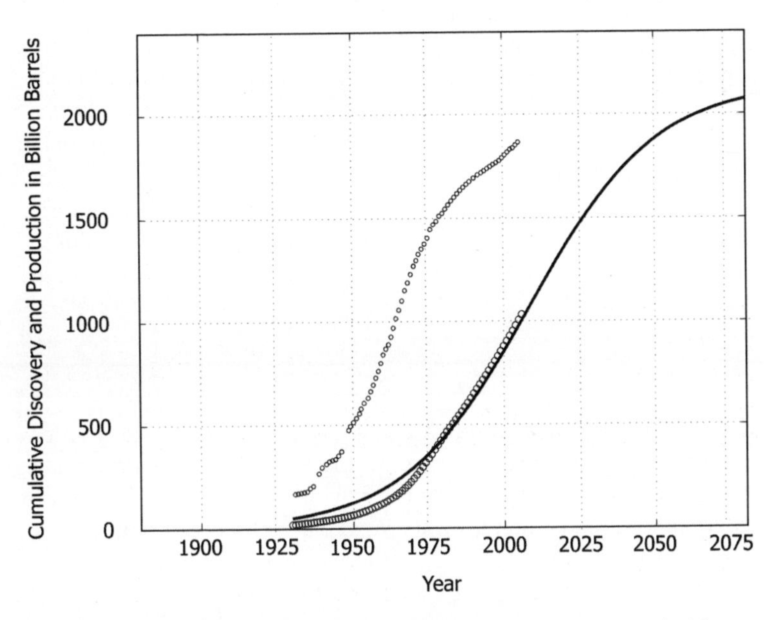

Figure 3.5 World cumulative oil discovery (left) and cumulative production (right)

Sources: (Left) J.H. Laherrère, "When Will Oil Production Decline Significantly?" European Geosciences Union, Vienna, Austria, April 3, 2006; (right) EIA-DOE and world pre-1970, C.J. Campbell, *The Coming Oil Crisis*, Multi-Science Publishing Company and Petroconsultants S.A., 1998.

as new data become available. These reflect any improvements in technology. Thus the Hubbert method, as applied here, looks forward and no claims are made regarding its predictive fitness for a long-term forecast. From this point of view it is similar to econometric models, which recalculate the performance going forward based on data on how the economy has performed in the past. When Hubbert's model is used with knowledge of backdated reserves, it enables better and better predictions to be made from each new yearly calculation.

Laherrère has improved on the Hubbert analysis by recognizing that many countries have more than one discovery cycle, as for example when offshore wells opened late. Each such cycle can be modeled with a logistic equation, and the production for the country then follows a sum of their individual contributions.

Economists tend towards a "consensus view" that, as prices rise, currently uneconomical oil will be produced, and reserves will increase through presently uneconomic resources being produced and consumed. Such an argument ignores the energy cost of production, and the question of energy return on energy invested (EROEI). When the energy for production is used in increasing quantities from the output itself, reducing net energy, clearly there comes a point at which the net energy delivered becomes zero. This will happen *long before* the world's difficult-to-produce (or "frontier," or "non-conventional") oil has been completely produced and consumed. Furthermore, oil is a remarkably useful fuel, as it exists in liquid form at atmospheric temperature and pressure, giving a high energy per unit of volume. Economists would have us believe that an equally good substitute will emerge to replace it, but as yet, nobody has identified one.

FUTURE

The modern world was built by coal, oil, and natural gas, which together account for 86 percent of primary energy production today. International commerce relied first on wind, then coal, then oil. With coal came railroads and railway barons. With oil came the automobile and the aeroplane. Oil pumped up international commerce to fabulous proportions. The "free trader" came to dominate the world. The result is a world insatiable for energy, now poised at the beginning of a protracted energy famine.

Adjusting to these new realities will be particularly difficult in countries such as the US, where life in sprawling suburbs is the rule, with greatly increased commuting. Oblivious to warning signs, people have spent the transient wealth made possible by concentrated energy on huge homes, heated by natural gas. With oil and natural gas entering their terminal decline, how these houses will be heated and how their inhabitants will commute to work is hard to imagine.

Economists still believe that economic growth will solve many of the problems facing humankind. They continue to ignore the predictable consequences of doubling demand for raw materials, as would be entailed by 3 percent population growth every generation.

Warnings have been coming thick and fast since the rise of coal, with Thomas Malthus's neither the first nor the last. Over half a century ago they were aired by Ayres and Scarlott, Putnam, Brown, and Cottrell. Now they are aired by Youngquist and Bartlett, among others. All have warned of the imminent collision course between growth in population, energy demand, and the limits to fossil fuels. Vogt, Osborn, Ehrenfeld, and Catton have sounded the alarm over the impact of growth on the entire ecological system. The Club of Rome brought these comprehensive messages to the reading public in their *Limits to Growth* over 30 years ago, with an update in 2004.[23]

With high-quality fuels on their way to exhaustion, governments will increasingly allow the public infrastructure to fall into disrepair. In proportion to their unwillingness to admit that geological limits have been reached for oil, followed by those for natural gas and coal, countries will try to defend the status quo, thereby worsening their situation. Modern societies cling fervently to a hope that biomass, wind, and solar energy sources will fill the gap left by fossil fuel depletion. Studies already show that the non-fossil fuels will only replace a fraction of fossil fuels. Fossil fuels represent a gift from nature from millions of years of geological processes, and they will all be used up by the few generations that occupied this planet in the late twentieth and early twenty-first centuries. Humanity is hopelessly trapped in a predicament unlike anything it has faced before.

APPENDIX

Hubbert's first method

The method used by M. King Hubbert to predict peak production for the lower 48 states of the US is based on the *logistic equation*. The same equation had been used by Pierre Verhulst to make calculations of human population growth in 1838. Only in 1980, by which time the US production peak had passed, did Hubbert give a full account of his methods.[24]

To understand the mathematical basis of Hubbert's method requires some knowledge of elementary calculus. Let Q denote the cumulative amount of oil that has been produced in some large oil province from the beginning of production to present. The logistic equation states that the rate of increase of cumulative production, Q', which can be taken to be the *annual production*, is given by the equation

$$Q' = aQ(1 - Q/Q_0)$$

Here, a is a parameter controlling the sharpness of the peak and Q_0 is the *ultimate production or ultimate recoverable reserves*. This is to say, the amount of oil that has been produced when the oil province is finally abandoned. Those familiar with calculus will recognize that Q' denotes the derivative of Q. Inspection of the right-hand side of this equation reveals a parabola, which increases from zero when Q is zero to a maximum, then drops to zero again at $Q = Q_0$. The maximum value is $aQ_0/4$, at the midpoint of ultimate production.

Before taking up the solution to the logistic equation, consider the early years of production. Then Q is much smaller than Q_0 and the logistic equation can be written in the approximate form

$$Q' = aQ$$

This is the same equation as used for calculating interest payments and sums, where Q represents the principal, which, when multiplied by the interest rate a, gives the yearly interest earned, Q'. For continuously compounded interest the familiar exponential growth formula is obtained.

For oil production cumulative production increases as

$$Q = Q_i \exp[a(t - t_i)]$$

Here Q_i is the cumulative production at time t_i, and exp denotes the exponential function, with t, the present year. The rapidity of the growth is determined by a, which is called the *intrinsic growth rate*.

A different and worthwhile view of the logistic equation is obtained by recasting it in terms of how much of the ultimate *remains to be produced*. If this is denoted by $Q_r = Q_0 - Q$, then toward the end of oil production the annual production can be determined from

$$Q_r' = -aQ_r$$

From this the remaining reserve, declines or decays as follows:

$$Q_r = Q_i \exp[-a(t - t_i)]$$

Thus the remaining oil decreases exponentially, and the parameter a, can now be called the *intrinsic decay rate*. Hubbert recognized that early in the production cycle, cumulative production increases exponentially and at the end it decreases exponentially. On this basis he drew by hand the possible production curve such that the area under his curve equaled the ultimate.

The cumulative production according to the logistic model can be shown to follow the formula

$$Q = Q_0 / (1 + \exp[-a(t - t_m)])$$

where t_m is the year of peak production, at which time cumulative production is $Q = Q_0/2$.

The three parameters in the solution are a, Q_0, and t_m. The most straightforward approach to estimate two of them is to recast the logistic equation into the form

$$Q'/Q = a - aQ/Q_0$$

and to interpret it as a straight line from coordinates showing Q on horizontal axis and Q'/Q on the vertical. The slope of this line is negative and has the value a/Q_0 It intersects the vertical axis at a, and the horizontal one at Q_0. Hence from this plot estimates can be made for both a, and Q_o, from the actual production history. The value of t_m can be determined from the known production record, which shows that at time t_i the cumulative production is Q_i. The only thing left to do is to decide which year and thus which pair of values to choose for t_i and Q_i.

For the United States the actual data is plotted in Figure 3.6 in the manner outlined above, with every tenth year indicated by a filled circle. Inspection of this graph shows that Hubbert's 1956 estimate

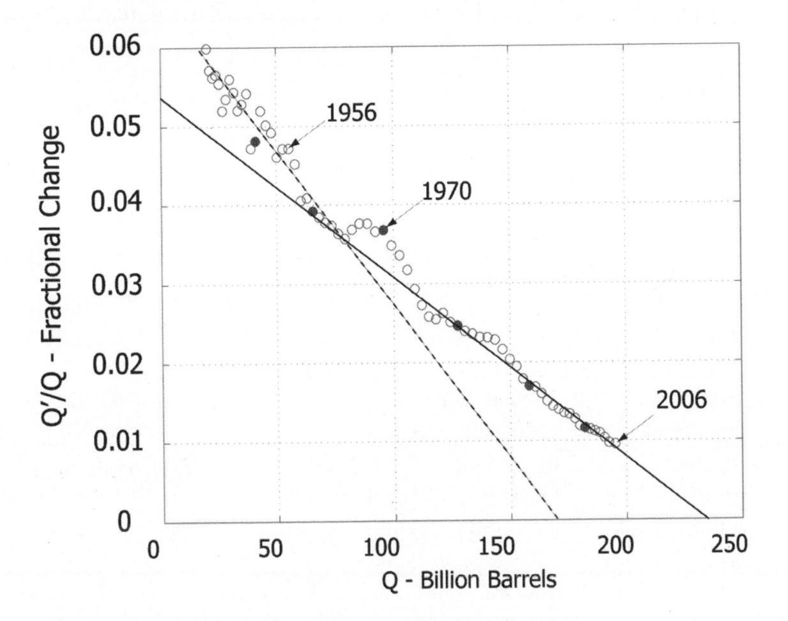

Figure 3.6 A plot to estimate the parameters for US oil production

Where Q is the cumulative production and Q' is the annual production.

Source: Author's calculations from data of previous figures in this chapter.

for the ultimate production, in other words the *initial oil endowment* of the lower 48 US oil, of between 150 and 200 Gb, is reasonable. The data in the early phase of a production cycle are subject to fluctuations, as yearly production is divided by the cumulative production, which remains small in the early phase. In 1962 Hubbert gave an estimate of 170 Gb, the best one could do with the data at hand. He reported an intrinsic growth factor of 0.067, when time is given in years, at this pre-peak phase. Today, fitting the last 17 years of data yields 240 Gb for the ultimate production and an intrinsic decay rate of 5.3 percent. The actual yearly decay rate has been 2.3 percent over the last ten years.

With the parameters a and Q_0 estimated, the yearly and cumulative production curves can be drawn. These are shown in the text for both the US and the world.

Hubbert's second method

The second method discussed next was described already in Hubbert's 1962 report, but its complete description appeared in 1982. In following this approach Hubbert rewrote the logistic equation in the form

$$N' + a\,N = 0, \text{ in which } N = Q_0/Q - 1$$

The solution of this equation is

$$N = N_i \exp(-at)$$

and N_i is the value of N at the beginning year, denoted as $t = 0$. Taking logarithms then gives the linear plot

$$\ln N = \ln N_i - at$$

The quantity Q_0 appears in both of the variables N and in N_i. By assuming a value for Q_0, say Q_b, the value for $N_b = Q_b/Q - 1$ is calculated each year of cumulated data on oil production. A least squares line then fitted through the last few years of the data.

Next define $b = Q_b/Q_0$, so that $N_b = bQ_0/Q - 1$, then if $b < 1$, as Q increases with each passing year toward Q_b, the value of N_b approaches zero at some time and thus its logarithm tends to minus infinity. That is, the curve $\ln N_b$ bends downward toward vertical. On the other hand, if $b > 1$, then as Q increases, it approaches Q_0 and N_b

tends to the positive value $b - 1$. This means that the curve flattens out toward horizontal. The correct value for Q_0 gives a straight line and its slope is the value of the parameter a, the intrinsic growth or decay rate in the logistic equation. The plots in Figure 3.7 illustrate the method.

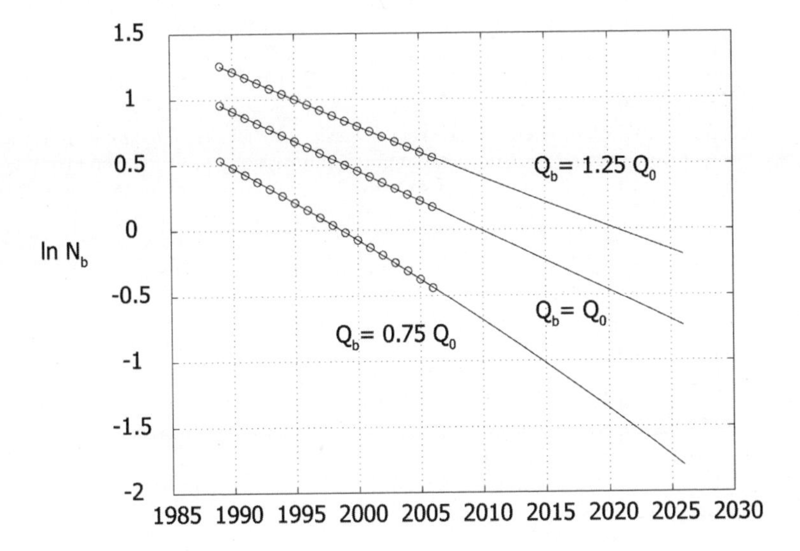

Figure 3.7 Construction to estimate a value for Q_0 for world oil

Source: Author's calculations from data of previous figures in this chapter.

With the parameters determined the curve can then be plotted. Using the last 17 years of data to construct a least squares fit, the world's ultimate production is estimated to be $Q_0 = 2{,}260$ Gb. This is slightly higher than the value $Q_0 = 2{,}200$ Gb obtained by the first method. The difference is about two years of consumption today.

According to the logistic model, the peak year corresponds to the time when half of the oil has been used. This means that, with best estimate, the value $N = Q_0/Q - 1 = 2 - 1 = 1$, and its logarithm is zero. The time of peak oil can be calculated from $t_p = t_i + \ln(Q_0/Q_i - 1)/a$.

That both methods give similar results is encouraging, for the first method could be criticized for the reason that the data points are not equally spaced, having maximum spacing at the midpoint of depletion. This is of little concern for predicting the peak production year. A more serious difficulty is toward the end of production when the points cluster together as Q' diminishes. What is gained by the second method is equally spaced data, but a slightly more complicated procedure to obtain a good value for Q_0.

NOTES

1. M. King Hubbert, "Energy from Fossil Fuels," *Science*, vol. 109, no. 2823 (1949); M. King Hubbert, "Nuclear Energy and the Fossil Fuels," presentation at the Spring Meeting of the Southern District Division of Production of the American Petroleum Institute, San Antonio, Texas, March 7–9, 1956.
2. Barbara Ward and René Dubois, *Only One Earth: The Care and Maintenance of a Small Planet.* An Unofficial Report Commissioned by the Secretary-General of the United Nations on the Human Environment, Prepared with the Assistance of a 152-Member Committee of Corresponding Consultants in 58 Countries. W.W. Norton, New York, 1972.
3. Eugene Ayres, "The Age of Fossil Fuels," in *Man's Role in Changing the Face of the Earth.* An International Symposium under the Co-chairmanship of Carl O. Sayer, Marston Bates, and Lewis Mumford. Ed. William L. Thomas Jr., University of Chicago Press, Chicago, 1956.
4. C.J. Campbell and J. Laherrère, "The End of Cheap Oil," *Scientific American* (March 1998).
5. C.J. Campbell, *The Coming Oil Crisis*, Multi-Science Publishing Company and Petroconsultants S.A., 1998.
6. C.J. Campbell, "Better Understanding Urged for Rapidly Depleting Reserves," *Oil & Gas Journal* (April 7, 1997).
7. C.J. Campbell, ASPO, www.peakoil.ie, accessed July 10, 2007.
8. J.H. Laherrère, "World Oil Supply, What Goes Up Must Come Down, but When Will it Peak?" *Oil & Gas Journal* (February 1, 1999).
9. L.F. Ivanhoe and G.G. Leckie, "Global Oil, Gas Fields, Sizes Tallied, Analyzed," *Oil & Gas Journal*, vol. 91, no. 7 (1993).
10. Walter Youngquist, private communication, August 7, 2007.
11. A.M.S. Bakhtiari, "IEA, OPEC Supply Forecast Challenged," *Oil & Gas Journal* (April 30, 2001); "OPEC Capacity Potential to Meet Projected Demand not Likely to Materialize," *Oil & Gas Journal* (July 9, 2001).
12. K.S. Deffeyes, *Hubbert's Peak*, Princeton University Press, Princeton, 2001.
13. *Oil & Gas Journal* (2006), EIA (2007) and IEA (2007).
14. EIA, *International Petroleum Monthly*, Table 1.4, http://www.eia.doe.gov/ipm/supply.html, accessed July 6, 2007.
15. Campbell, *The Coming Oil Crisis.*
16. *Oil & Gas Journal*, "Oil Production, Reserves Increase Slightly in 2006" (December 18, 2006).
17. US Geological Survey World Energy Assessment Team, US Geological Survey World Energy Petroleum Assessment 2000, http://energy.cr.usgs.gov/WEcont/world/woutsum.pdf, accessed November 11, 2002.
18. M. King Hubbert, "Techniques of Prediction as Applied to the Production of Oil and Gas." Presented to a symposium of the US Department of Commerce, Washington, June 18–20, 1980. In Saul I. Gass, ed., *Oil*

and Gas Supply Modeling, Bureau of Standards special publication 631, Washington, National Bureau of Standards, 1982, pp. 16–142.

19. EIA, "International Energy Outlook 2007" (May 2007).
20. P. McCabe, "Energy Resources – Cornucopia or an Empty Barrel?" *Bulletin of the American Association of Petroleum Geologists*, vol. 82 (1998), pp. 2110–34.
21. S.A. Korpela, "Oil Depletion in the World," *Current Science*, vol. 91, no. 9 (November 10, 2005).
22. J.H. Laherrère, "When Will Oil Production Decline Significantly?" European Geosciences Union, Vienna, Austria, April 3, 2006.
23. Eugene Ayres and Charles A. Scarlott, *Energy Sources: The Wealth of the World*, McGraw-Hill, New York, 1952; Palmer C. Putnam, *Energy in the Future*, Van Nostrand, New York, 1953; Harrison Brown, *The Challenge of Man's Future*, Viking Press, New York, 1954; Fred Cottrell, *Energy and Society*, McGraw-Hill, New York, 1955; W. Youngquist, *Geodestinies*, National Book Company, Portland, Oregon, 1997; A. Bartlett, "An Analysis of US and World Oil Production Patterns Using Hubbert-style Curves," *Mathematical Geology*, vol. 32, no. 1. (2000); A. Bartlett, "Arithmetic, Population and Energy," www.hawaii.gov/dbedt/ert/symposium/bartlett/bartlett1.html; David Ehrenfeld, *The Arrogance of Humanism*, Oxford University Press, Oxford, 1972; William Catton Jr., *Overshoot*, University of Illinois Press, Illinois, 1980; Donella H. Meadows, Dennis L. Meadows, Jørgen Randers, and William W. Behrens III, *The Limits to Growth*, Universe Books, New York, 1972; Donella H. Meadows, Jørgen Randers, and Dennis Meadows, *Limits to Growth, The 30-Year Update*, Chelsea Green Publishing, White River Junction, Vermont, 2004; William Vogt, *Road to Survival*, William Sloane Assoc., New York, 1948; Fairfield Osborn, *Our Plundered Planet*, Pyramid Books, Little, Brown, and Co., Boston, 1948.
24. *Hubbert*, "Techniques of Prediction as Applied to the Production of Oil and Gas"; M.K. Hubbert, *Energy Resources: A Report to the Committee on Natural Resources*, National Academy of Sciences, National Research Council, Publication 1000-D, 1962.

4
The Assessment and Importance of Oil Depletion

Colin J. Campbell

Oil provides some 40 percent of the world's energy needs and as much as 90 percent of its transport fuel. It also has a critical role in agriculture, which provides food for the world's population of 6 billion people. It is, however, a finite commodity, having been formed in the geological past, which means that it is subject to depletion. Given that it is of such great importance to the modern world, it is indeed surprising that more attention has not been given to determining the status of depletion.

There are several possible explanations for this strange state of affairs. First, it is counter-intuitive. The weekly trip to the filling station is such a normal part of daily life that most people see a continued supply of oil as being as much a part of nature as are the rivers that flow from the mountains to the sea. Second, depletion is strangely foreign to classical economics, which depict man as the master of his environment under ineluctable laws of supply and demand. Never before have resource constraints of such a critical commodity begun to appear without sign of a better substitute or market signals. The reason for the absence of early market warning is due to expropriations that have obscured the natural trends that would otherwise have alerted us to growing shortages and rising costs. Tax by both consuming and producing countries has furthermore distorted the position. A related issue is a blind faith in technology, as epitomized by the dictum "the scientists will think of something." Unfortunately, if they do, they will simply deplete the remaining oil faster. Third is the denial and obfuscation by the oil industry, which is in a position to understand the situation but finds itself the victim of an investment community driven by imagery and a very short-term view of the future. The industry is itself subject to internal vested interests, represented, for example, by the explorers, whose careers are not served by pointing out the natural limits. The oil companies can accordingly be excused for choosing their words with such extreme care. Fourth is the nature of so-called democratic government that finds it easier to attract votes by reacting to crises than by anticipating them, especially where unpopular or even draconian responses are called for. Fifth are possible conspiracies by countries already dependent on rising oil imports, which seek to secure access to supply and hold prices down by any means at their disposal.

There may be other factors at work too, but these five elements offer a range of possible explanations for why the subject is so clouded by mystery and disbelief. In strictly technical terms, there is nothing particularly difficult in assessing the size of an oilfield or in extrapolating the discovery trend to indicate what remains to be found in the future. I will try here to lift the veils, to provide a fair statement of the true position.

BACKGROUND

Oil from seepages has been known since biblical times, being used, for example, as mortar in Babylon, but the modern oil industry had its roots in the nineteenth century, when it commenced drilling for oil on the shores of the Caspian and in Pennsylvania. The technology was not new, as wells had been drilled earlier to tap salt brines, needed to preserve meat in the days before refrigeration. The science of petroleum geology evolved rapidly to provide a technical basis for exploration. It was soon appreciated that oil resulted from the decomposition of microscopic organisms and that, once formed, it could be trapped in certain geological structures, which could be identified and mapped.

Attention turned to the Middle East during the early years of the last century, prompted first by seepages in the Zagros Foothills of Iraq and Iran, but later by the less obvious prospects beneath the sands of Kuwait and Saudi Arabia, which were identified only with the help of core-drilling and seismic surveys. The prolific oil lands of the United States, Mexico, Venezuela, and Indonesia were also opened up in parallel. Exploration expanded throughout the world, so that most of the onshore oil basins and many of the giant fields within them had been identified prior to World War II.

The demand for oil grew rapidly from the economic boom that followed the war, prompting further exploration, which led to important finds in the Soviet Union and in Africa, as well as the opening of offshore drilling, which was greatly facilitated by the development of the semi-submersible rig in 1962. Floating on submerged pontoons beneath the wave base, it brought routine drilling to the continental shelves of the world.

From the earliest days, the oil industry has been characterized by "boom and bust" cycles, for the simple reason that, once found, oil flows from the ground at great pressure, which contrasts with the painfully slow extraction of coal and minerals by pick and shovel. Prolific production from new finds flooded the market and depressed the price, which in turn inhibited new exploration until the early wells began to run dry, when the cycle was repeated by new discovery in new areas. It became evident that some control of the open market was called for to avoid these damaging fluctuations. The US was the first to apply it, using the Texas Railroad Commission to maintain price by rationing production. Its example was followed by the creation of the Organization of Petroleum Exporting Countries (OPEC) in 1960, which sought to perform the same much needed function on a world scale.

The wealth from oil, known as black gold, became legendary. The industry grew to be the largest in the world, and many of the producing countries began to rely heavily on oil. As their economies and populations grew, so did their appetite for oil revenues, which in turn led to expropriations, prompted also by the belief that they were not being fairly compensated for the depletion of their natural resources. The Soviet Union was the first to move, expropriating the assets of the foreign companies in 1928, followed ten years later by Mexico. The trend accelerated after the war, starting in Iran in 1951, and extending during the 1970s to the other main producers – Iraq, Kuwait, Saudi Arabia, Libya, Venezuela, and Algeria. But it did not quite deliver the anticipated benefits. The international companies had been previously able to take the revenues paid to the foreign producing countries as a charge against home country tax. The producing countries lost this hidden subsidy following expropriation when they had to face the raw pressures of the open market. Prices became volatile as a consequence.

In the period 1947–49, Britain, which had administered the territory of Palestine as a protectorate, surrendered to terrorist pressures. Massive Jewish immigration led to conflicts and the occupation of new lands, forcing the indigenous Arabs into refugee camps. Tensions mounted in 1973, causing certain sympathetic Arab oil producers to restrict exports to the US and the Netherlands, which were perceived to side with Israel in the conflict. Although it was a short-lived restriction of only a few months, oil prices rose five-fold in what became known as the first oil shock, which plunged the world into recession, curbing oil demand. It was, in turn, followed five years later by a second price shock, occasioned by panic buying following the fall of the Shah of Iran, when oil prices soared to the equivalent of US$80–100 a barrel, in today's money.

These two oil price shocks, while themselves transitory and politically motivated, demonstrated the degree to which the world had become dependent on cheap oil. That in turn prompted several contemporaneous studies of depletion, epitomized by the well-known report "The Limits to Growth," which in fact, like "Blueprint for Survival" (from *The Ecologist*) was published just before the first oil shock.

The resources themselves were far from running out, but it was realized that production would eventually reach a peak and decline. This obvious conclusion was not exactly new; in 1956 M. King Hubbert had already drawn a simple bell curve showing that US production would peak in 1971, at the midpoint of depletion. He could readily estimate the total endowment below the curve, as discovery had peaked 40 years before the study. Since oil has to be found before it can be produced, it is obvious that production has to mirror discovery, after a time-lag.

But the warning signals were both ignored and misrepresented, as new production from Alaska and the offshore, including particularly the prolific North Sea, flooded the world with cheap oil. Having lost their principal sources of supply through expropriation, the international companies concentrated on these new areas, which they did control, and worked flat-out. In Europe, socialist governments, which could have managed their countries' resources in the national interest through state companies, were replaced by doctrinaire free marketers, epitomized by Mrs. Thatcher, who encouraged the rapid depletion of Britain's oil and gas. The North Sea has now peaked, and is declining at about 6 percent a year, meaning that production will have roughly halved in ten years' time (see Figure 4.1).

In 1981 the rate of discovery began to fall short of consumption, despite a surge of tax-driven drilling, and the deficit has grown ever since (see Figure 4.2). Although the international companies continued to speak optimistically about their future growth, their actions told a different story. By the end of the century, the so-called Seven Sisters, which had dominated the world of oil for so long, had been reduced to just four: Shell in splendid isolation, Exxon

Mobil, BP Amoco Arco, and Chevron Texaco, with a second tier in Europe comprised of Total, Elf, and Fina. The major companies began to merge, downsize and shed staff. One changed its logo to a sunburst and claimed that its initials stood for "Beyond Petroleum," as a very oblique reference to the depletion of its principal asset. The so-called independent companies, too, were disappearing through merger and acquisition, as was the contracting business on which they all depend.

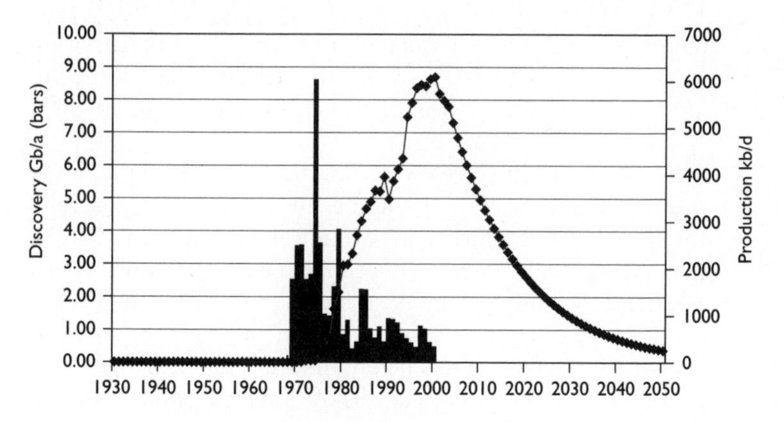

Figure 4.1 Oil discovery and production in the North Sea

Source: This figure, and all subsequent figures in this chapter, compiled by the author from various sources and industry data.

The production of conventional crude oil outside the five main Middle East producers reached a peak in 1997, but falling demand from an Asian recession combined with increasing Russian exports made possible by the weak ruble, led to an anomalous fall in oil prices. That was, in turn, followed by the reappearance of supply capacity limits which caused prices to triple – that is, a 200 percent increase in price – during the latter part of 1999. The high price was believed, in certain financial milieux, to have been a cause of the recession that started in 2000, which weakened demand, reducing pressure on oil prices in a vicious circle likely to be repeated in the years ahead.

The last chapter of this unfolding saga came on September 11, 2001. The action was soon attributed to a Saudi dissident living in a cave in Afghanistan, and prompted the US to declare a global war on terrorism. Afghanistan was bombed, toppling its Taliban

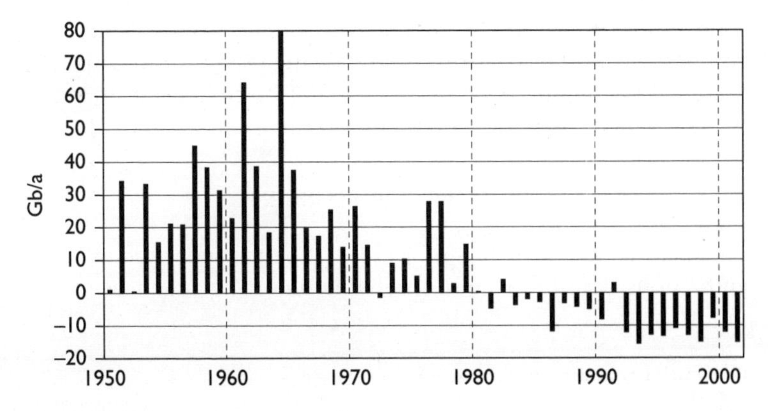

Figure 4.2 Discovery–consumption gap

government, with which the US government had no difficulties in dealing – for example, on pipeline routes through Afghanistan, and within months before September 11. During Soviet occupation, US governments had actively supported and financed Osama bin Laden. Israel took the opportunity to step up the brutal suppression of its indigenous population, claiming common cause with the US in a war on terrorism. A brief unattributed anthrax scare in the US mobilized public opinion, bringing unprecedented popularity to President Bush, who declared several oil-producing countries to be an "Axis of Evil" – the term "axis" having associations with the perpetrators of the Holocaust. He threatened to invade oil-rich Iraq, and started stockpiling military fuel supplies for the purpose. US military bases were established around the Caspian oilfields, while in Latin America the US was implicated in a failed plot to overthrow the President of Venezuela, the strongman of OPEC. Observers can be forgiven for concluding that the new foreign policy of the United States has a hidden oil agenda.

What may follow from this chain of events is impossible to predict, but it is at least on the cards that popular outrage in the Middle East, prompted by US military intervention, may lead to the fall of the Saudi government, giving the US the pretext to take the Arabian oilfields by force. The US has long explicitly declared that it regards access to foreign oil as a vital national interest, justifying military intervention where necessary. As its own domestic production continues to decline without hope of reprieve, its need for foreign oil becomes ever more desperate. Whereas in the past military intervention may have been contemplated as a reaction to politically motivated interruptions to supply, now it faces the inevitable consequences of depletion and conflict, with other countries also seeking a share of what is left. Much of that lies in the Middle East, thanks to circumstances in the Jurassic period over which no politician can exercise control.

MISLEADING OIL REPORTING

The main reasons why this subject is not better understood are the ambiguous definitions and unreliable reporting practices of the industry.

Conventional and non-conventional oil

Oil is oil from the standpoint of the motorist filling his tank, who does not much care from whence it comes. But the analyst of depletion needs to identify the different categories, because each has its own costs, characteristics, and extraction rates, and hence can contribute differently to peak production. The term *conventional* is widely used to describe the traditional sources, which have contributed most oil produced to date, and which will dominate all supply far into the future.

There are, in addition, *non-conventional* sources, which will be increasingly important when conventional oil declines after peak, but there is no standard definition of the boundary. Here, the following categories are treated as non-conventional, and are described in greater detail in a later section.

Oil from coal and "shale" (actually immature source-rock)
Bitumen
Extra-heavy oil (density < 10° API)[1]
Heavy oil (density 10–17.5° API)
Polar oil and gas
Deepwater oil and gas (> 500 m water depth)

Liquids that condense naturally from the gas caps of oilfields are included with crude oil, but the liquids extracted from gas by processing are treated separately.

Production and supply reporting

Measuring production is simply a matter of reading the meter, but national statistics are confused by the inconsistent treatment of natural gas liquids, war loss (which is production at least in a technical sense) and frontier changes. Supply is not the same as production, but includes stock change and refinery gains. The reporting of gas production is still more confused, referring variously to raw gas or marketed gas after the removal of inert gases such as nitrogen and carbon dioxide, which are often present, with the differing treatment of flared and re-injected gas adding to the uncertainty.

Reserve reporting

The practices of reserve reporting evolved early, being much influenced by the environment of the old onshore fields of the US, which were characterized by a highly fragmented ownership. Being onshore and close to market, the wells could be placed on production as soon as they had been completed. To prevent fraud, the Securities and Exchange Commission (SEC) introduced strict rules whereby owners could treat as proved for financial purposes only the reserves "behind pipe," meaning those to be drained by existing wells. As the fields were drilled up, the reported reserves naturally grew. In practice, the reserves of the old fields were mainly estimated by extrapolating the decline rates of the wells. The highly fragmented ownership meant that there was little interest in, or indeed possibility of, making field-wide reserve estimates. The companies did, however, recognize additional probable and possible reserves that did not qualify for proved financial status.

Different conditions obtained overseas and in offshore areas, where it was normal for fields to be developed as single entities by one or more companies acting as a group. They were more interested in what the field as a whole would produce over its full life, especially offshore, where they had to design appropriate facilities in advance of production. They were still lumbered with the SEC rules, which required the reporting of proved reserves, although in practice they reported better estimates of what the fields would deliver over their full lives. There is naturally a degree of latitude in estimating future production. For a variety of commercial reasons, it was found expedient to report ultra-conservative estimates of discovery, which consequently grew over time, delivering an attractive impression of gradually appreciating assets to the stock market, and serving to reduce tax in countries operating a depletion allowance. Many of the large North Sea fields, for example, were initially under-reported by about one-third. This luxury is not however available to the more recent small fields, with a short life and high economic threshold.

They in fact sometimes give disappointing results, yielding negative reserve growth.

A further confusion has arisen from the application of probability theory, in which reserves are equated against differing subjective probability rankings. Under this system, *proved reserves (1P)* are commonly equated with a 95 percent probability, whereas *proved & probable & possible reserves (3P)* are held to have a 5 percent probability. *Mean, median,* and *mode* values are then computed. This system appeals to the scientifically inclined, but in practice adds to the confusion. In plain language, *proved reserves* relate to the current status of development, whereas *proved & probable reserves* (or *"mean"* under the probability system) are estimates of what the field as a whole is expected to produce over the rest of its life. The probability range is unnecessarily wide, as engineers with modern methods can make good estimates. Why should anybody be interested in an assessment having no more than a 5 percent probability of being correct, and a subjective one at that?

The dating of reserves revisions

For financial purposes, reserve revisions are reported on the date that they are made, but this gives a misleading impression of the discovery trend. Year-on-year comparison of national reserves, with subtraction of the intervening production, gives the impression that more is being found than is the case, which has misled many analysts working with data in the public domain.

To determine a valid discovery trend, it is necessary first to make sure that the reported production and proved & probable reserves relate to the same categories of oil; and second to backdate any revisions to the discovery of respective fields. In practice, this cannot be done without access to the industry database to identify the details, and its cost puts it out of range for most analysts.

Several OPEC countries announced colossal overnight reserve increases in the late 1980s, when they were vying for quotas based on reserves. While some upward revision was called for, as the earlier numbers were too conservative, having been inherited from the private companies before they were expropriated, the revisions had to be backdated to the discovery of the fields containing them, some of which had been found up to 50 years earlier.

Dating the reserves is as important, if not more so, than estimating the amounts. The explanation for the revisions is another highly important matter to grasp. The industry, not wishing to admit to poor reporting practices, has found it expedient to attribute revisions to technological progress when they were in fact mainly a reporting phenomenon. This in turn carries the danger of unjustifiably extrapolating reserve growth into the future on the assumption of an inexorable march of technological progress. No one disputes the progress to date, but its main impact has been to hold production higher for longer, which makes good economic sense but accelerates depletion.

Data sources

Two trade journals, the *Oil & Gas Journal* and *World Oil*, have compiled information on production and reserves for many years on the basis of a questionnaire sent out to governments and others. Many of the reports remain implausibly unchanged for years on end, simply because the country concerned has failed to update its estimate, despite production. There are also substantial discrepancies between the two data sets despite the fact that they are compiled in a similar fashion.

A third source is the *BP Statistical Review of World Energy*, which is the most misleading of all, because many analysts wrongly assume that the reported oil reserves have at least the tacit blessing of a competent and knowledgeable oil company in a position to assess their validity. In fact, BP simply reproduces the *Oil & Gas Journal* oil reserve data, save in one or two specific cases.

These public sources contain information very different from that in the industry's own database, which is compiled on a field-by-field basis directly from the companies' own records. This itself contains certain anomalies, and seems to be deteriorating in quality as it faces the increasingly difficult challenge of compiling information from the proliferation of small companies and ever less reliable state information. Particular difficulties are faced in interpreting data from the former Soviet Union, which operated its own system of reserve classification that tended to ignore economic constraints.

In short, although there are no particular technical difficulties in estimating the size of an oilfield, especially with the advantage of modern technology, the reporting of production and reserves remains highly unreliable. In these circumstances it is well to confirm, wherever possible, the estimates of individual fields by extrapolating the decline, which plots as a straight line on a graph relating annual to cumulative production (see Figure 4.3).

ESTIMATING FUTURE DISCOVERY

In earlier years, it was difficult to identify the precise source of the oil that found its way into oilfields, but a geochemical breakthrough in the 1980s resolved the issue. Isotopic examinations showed that oil was derived from algae (and similar micro-organisms), whereas gas came from vegetal material, as well as deeply buried oil, which had been broken down into gas by high temperatures. This knowledge led in turn to the realization that the bulk of the world's oil came from no more than a few epochs of extreme global warming, when prolific algal growths effectively poisoned the seas and lakes. Petroleum geology made great advances, making it possible to map the world's producing belts once the critical data had been gathered from seismic surveys and preliminary boreholes. The world has now been so extensively explored that virtually all the productive belts have been identified, save perhaps in certain polar and deepwater regions, which are here treated as non-conventional, partly for that very reason.

Estimating the future discovery of an established basin is a straightforward task, achieved by extrapolating the discovery trend with a so-called *creaming curve*, which plots cumulative discovery against cumulative *wildcats* (exploration boreholes), and by studying field size distributions with a *parabolic fractal*. The larger fields are generally found first, for the simple reason that they are difficult to miss,

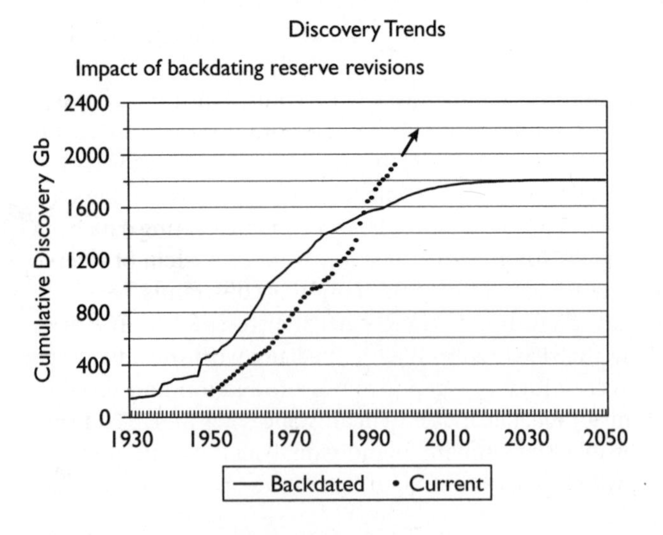

Figure 4.3 Annual vs. cumulative production

being followed in turn by progressively smaller finds (see Figures 4.4 and 4.5).

A FLAWED STUDY BY THE UNITED STATES GEOLOGICAL SURVEY

The USGS started evaluating the world's oil resources following the oil shocks of the 1970s, and under its previous director put out sound evaluations that were published at successive World Petroleum Congresses. A departure was issued in 2000, which greatly exaggerated the scope for new discovery and the "growth" of existing reserves. It is worth briefly commenting on this flawed study because it has misled several foreign governments and agencies, including the International Energy Agency.

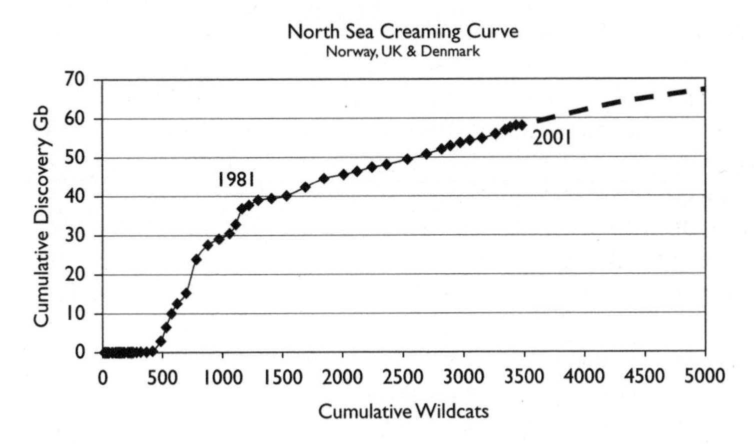

Figure 4.4 North Sea creaming curve

The study started by usefully identifying all the prospective basins of the world. It did not extrapolate past discovery with the methods outlined above, but relied on abstract geological assessment, subject to a range of subjective probability rankings. Thus for example, in the case of an undrilled basin in East Greenland, it determined that there was a 95 percent chance (F95) of it containing more than zero, namely at least one barrel, and a 5 percent chance (F5) of it containing more than 112 giga-barrels (Gb), from which a mean value of 47 billion was computed.

In reality, the 5 percent chance cases cannot be other than wild guesses that could as well give half or double the true value, yet they influenced the computation of mean values, which were summed to

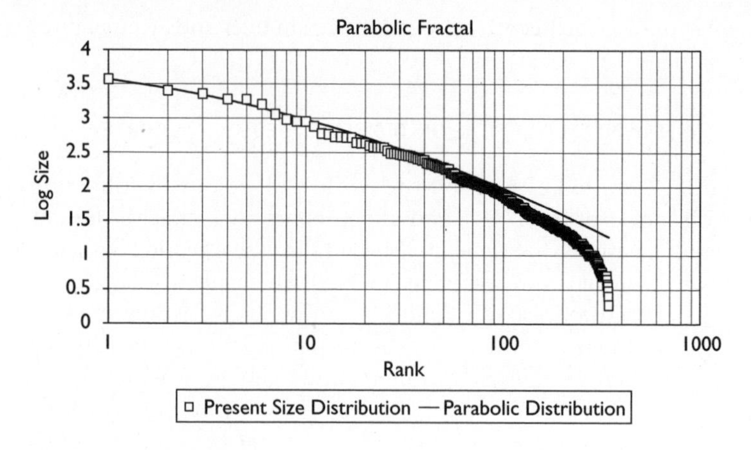

Figure 4.5 Parabolic fractal

give the world total. The indicated amounts related to discovery over a 30-year period starting in 1995. While there may be some abstract scientific merit to the study, it says little about what will actually be found in the real world, as is well confirmed by the results to date. The mean estimates imply an average discovery of 25 Gb a year, when so far the actual average has been only 10 Gb, which is doubly damning because above-average results are to be expected during the early years, as the larger fields are normally found first.

The estimates of "reserve growth" are equally flawed, being based on the experience of the old onshore US fields, which, as discussed above, are not remotely representative of the offshore or overseas fields. To its credit, the USGS did express serious reservations about the estimations in the accompanying text. While the study itself speaks of academic inexperience and ineptitude, political overtones were introduced when the USGS issued a press release of the unfinished study on the eve of a critical OPEC meeting, and by the fact that it goes to great lengths to publicize the study at conferences around the world and by direct interventions with foreign governments and agencies.

It is, at the same time, curious to find a member of the USGS team publishing an impressive poster that depicts the imminent peak of oil production, termed the great Roll-Over, with a text speaking of a rough ride if the world does not wake up to the reality of its predicament.

ESTIMATING FUTURE PRODUCTION

If we had reliable production and reserve data, and it is a very big *if*, it would be a fairly straightforward task to forecast future production using one or more of the following statistical techniques:

1. ***Simple depletion***. The simplest, and in some ways perhaps the best model, is to divide the world into three groups:

   - countries past their depletion midpoint, where production is expected to decline at the current depletion rate (annual production as a percentage of total future production);
   - countries that have not yet reached their midpoint, whose production is set to continue to rise until midpoint, before declining at the then depletion rate;
   - swing countries, comprising the five major Middle East producers (Abu Dhabi, Iran, Iraq, Kuwait, and Saudi Arabia), which make up the difference between world demand under various scenarios and what the other countries can produce under the model. The current base case scenario is that conventional production will be on average flat as a result of alternating price shocks and consequential recessions until 2010, when the swing producers can in practice no longer offset the declines elsewhere, and world production commences its terminal decline at the then depletion rate.

   This method is used in the ASPO Statistical Review. It has the advantage of recognizing demand impacts, not easily covered in the strictly statistical methods described below.
2. ***Hubbert models***. Production in an unfettered environment can be modeled with a simple Hubbert bell curve based on an estimate of ultimate recovery, or with multiple curves reflecting different cycles of discovery and corresponding production. In world terms, a simple Hubbert curve, built on the indicated size of the resource (1,900 Gb), shows a peak in 1995 at 40 million barrels/day, but was not realized because the oil shocks of the 1970s curbed demand, giving a lower and later peak (see Figure 4.6).
3. ***Discovery–production correlation***. Since production has to mirror earlier discovery, future production can be modeled by superimposing the production trend on the past discovery trend with a time shift, as demonstrated by Laherrère.

4. ***Rate plots***. There is a mathematical procedure that converts a bell curve into a straight line, achieved by plotting annual production as a percentage of cumulative production on one axis, against cumulative production on the other. The straight line can be readily extrapolated, as explained by Deffeyes.

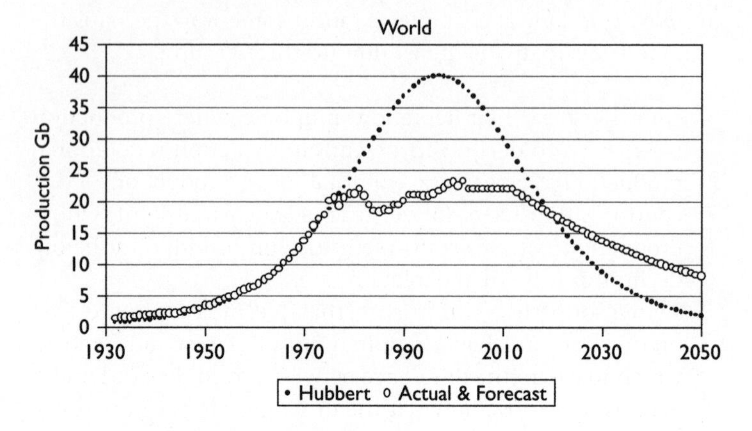

Figure 4.6 Hubbert curve

GAS

Gas is more difficult to evaluate than oil. Some occurs in discrete deposits, known as dry gas, being mainly derived from deeply buried coals, but most occurs in the gas caps of oilfields, known as associated gas. About 80 percent of the gas in a reservoir is recoverable, compared with only about 40 percent for oil. Liquid hydrocarbons, known as *condensate*, condense naturally from gas on being brought to the surface, and more may be extracted by processing, both forming important resources for the future.

Gas depletes very differently from oil due to its higher mobility. An uncontrolled well would deplete a gas deposit quickly, and production is normally deliberately capped far below the natural capacity, commonly by the simple expedient of pipeline pressure. Accordingly, production normally follows a long plateau, with most fluctuation being seasonal. Gas prices generally fall as the investment costs are written off, which in turn attracts new customers. Production continues along the set plateau for a long time, but when the inbuilt spare capacity has been drawn down, it comes to an abrupt end without many market signals. Whereas oil trade is global, the gas market is regional, built around the hubs of North America, Europe, and the Far East. The US faces a crippling shortage of gas as it reaches the end of its plateau of production, and will rely increasingly on imports from Canada and the Arctic. Europe is better placed, being able to draw on supplies from North Africa, Russia, Central Asia, and eventually the Middle East, after its own North Sea production ends. The Far East and China have to depend on local sources, as well as imports from Russia and Central Asia. The abrupt end of gas production needs to be recognized by the responsible authorities, because the market delivers no warning signals.

Modeling gas supply, with its hidden inbuilt spare capacity, is difficult, as so much depends on infrastructures and markets. Here, global production is expected to rise to a plateau of 170 trillion cubic feet per annum, lasting from 2015 to 2040, but the forecast is most uncertain.

NON-CONVENTIONAL OIL AND GAS

Heavy oils

Oil may be extracted from coal by the use of the Fischer Tropf process, invented in Germany during World War II, and it may be retorted from immature oil source rocks, termed "oil shales." Much interest was shown in the latter method after the oil shocks of the 1970s, but all projects came to naught. The residue is a fine, toxic powder carrying environmental hazards and costs, and the net energy return is very poor.

Conventional oil migrated to the margins of the basins of western Canada and eastern Venezuela in substantial quantities, where it was weathered and attacked by bacteria. The light fractions were removed, leaving behind bitumen (defined by viscosity) grading into extra-heavy oil (defined by density). In Canada, the so-called tar-sands containing the bitumen are mined at the surface after the removal of up to 75 meters of overburden. The ore, for that is what it is, is centrifuged and processed in plants fueled by cheap stranded gas to yield a light, high-quality synthetic oil. In Venezuela the deposits lie at 500- to 1,500-meter depths and are produced with the help of steam injection from closely spaced wells. The extra-heavy oil grades into heavy oil, which is here arbitrarily defined as that denser than 17.5 API. There are other deposits around the world, but those of Canada and Venezuela are the most important.

The resources are enormous, but the extraction rate is low and costly. No doubt production will be stepped up from the current level of about 2 million barrels/day after the peak of conventional oil, but it is difficult to imagine it exceeding about 5 million barrels/day by 2020, despite superhuman effort and every financial incentive. It is also worth remembering that the deposits are not homogeneous: even a small addition in the thickness of overburden adds greatly to the cost of tar-sand extraction. Processing also uses fuel, which will become increasingly expensive once the stranded gas deposits, currently used, have been exhausted.

Deepwater oil and gas

In earlier years, the deepwater domain was considered too far from land to contain oil reservoirs and source rocks. But recent exploration has identified certain areas in divergent plate settings where Cretaceous rifts yield source rocks, and where turbidity currents comparable with submarine avalanches brought in sands to form reservoirs, especially where winnowed by long shore currents. These special conditions appear to be restricted to the Gulf of Mexico and the margins of the South Atlantic. Deltas elsewhere may locally extend into deep water, but lacking underlying prolific source-rock, any petroliferous potential they might have will rely upon whatever source-rocks occur within the delta itself, which are likely to be gas-prone. Present evidence points to a total endowment of about 65 Gb. If all goes well, production may peak at around 8 million barrels/day within a year or two of 2010. It is axiomatic that no one would look for oil under these extreme conditions if there were anywhere else easier left.

Polar oil and gas

Antarctica appears to have very limited geological prospects, and is in any case closed to exploration by agreement. The Arctic regions are more promising, although large vertical movements of the crust under the weight of fluctuating ice caps in the geological past have tended to depress the source-rocks into the gas window. Alaska is an exception, but appears to be a concentrated habitat with most of its oil in the giant Prudhoe Bay field, which has been in decline since 1989. There are also substantial oil deposits in the Siberian Arctic, here tentatively estimated at 30 Gb, with production reaching a peak of about 5.5 million barrels/day by 2020. The gas reserves throughout the Arctic are likely to be very large indeed, but extraction will be slow and costly in this extreme environment.

Non-conventional gases

Coalbed methane, derived from coal deposits, is an important non-conventional gas already supplying about 6 percent of US needs. More can be expected from the other coal-bearing regions of the world. Another useful source is gas extracted from fractures in hydrocarbon source-rocks, known as "tight gas." Much attention has been given to gas hydrates, which, it has been wrongly claimed, form large deposits in deepwater and polar regions. In reality, the methane occurs in disseminated granules and laminae, which are unlikely to be producible.

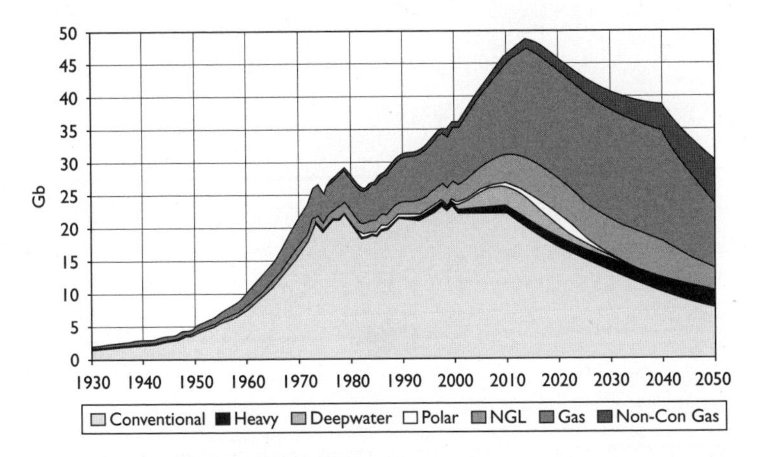

Figure 4.7 Depletion of all hydrocarbons

Figure 4.7 depicts the depletion of all hydrocarbons as modeled herein. It will be noted that the production of conventional oil is expected to be about flat until 2010, due to alternating price shocks and consequential recessions dampening demand. The peak of all liquids also comes around 2010, with gas following about 15 years later. It means that the production of all liquids need not fall below present levels for about 20 years, assuming that the deepwater and polar oil, and natural gas liquids, come in as expected. In the unlikely event that sustained economic growth could be restored, demand would rise accordingly, advancing the peak and steepening the ensuing decline.

WORLD REGIONAL ASSESSMENTS

United States

Discovery in the US peaked in 1930, followed 40 years later by the corresponding peak in production. Alaska provided a secondary cycle, but was insufficient to reverse the decline, and the new deepwater Gulf of Mexico offers a third (see Figure 4.8). It is doubted if the Alaska Natural Wildlife Refuge (ANWR) area, which is closed for environmental reasons after the drilling of one very confidential borehole, would make any material difference if opened. US oil imports already run at about US$130 billion a year and are set to rise unless the government can somehow introduce draconian policies to cut demand. Its gas supply is even more critical, as already discussed. It is hard to avoid the conclusion that this looming energy crisis will spell the end of the American dream and US global economic hegemony, even if the country goes down with all guns blazing.

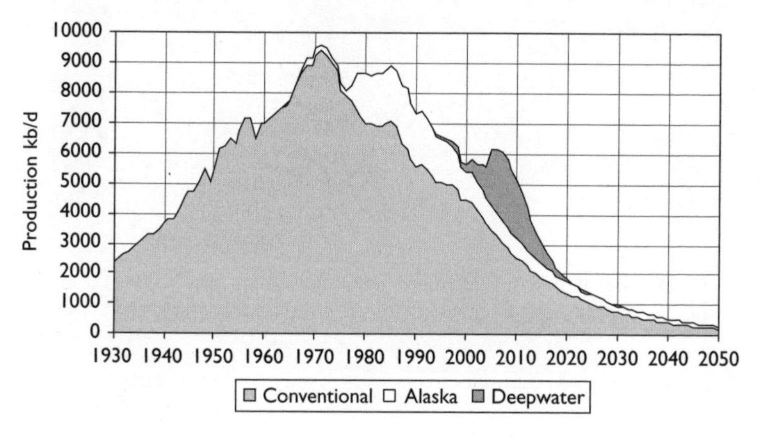

Figure 4.8 US production

Russia

Russian discovery peaked in the 1960s, followed in 1987 by a peak in production at just over 11 million barrels/day. Production, which fell precipitately on the collapse of the Soviet government, is now set to increase to a second slightly lower peak around 2010, in part bringing in what would have already been produced but for the interruption. In addition there may be substantial production of non-conventional oil from the Arctic (see Figure 4.9). Russia's gas deposits are very large indeed, with reserves amounting to about one-third of the world's total, which will see increasingly high demand from Europe, China, and the Orient.

Russia is experiencing an epoch of hyper-capitalism, with the emergence of various oil barons who could put Mr. Rockefeller to shame. At the present time they are bent on exporting at the maximum rate possible to earn foreign exchange, and are able to undercut world prices thanks to the devalued ruble, which holds down their operating costs. It is entirely possible that Russia may take over from the Middle East the role of swing oil producer during the next decade or so, and it is already building a dominant position in Eastern Hemisphere gas markets. This confers great geopolitical strength, and responsibility both for the country itself and the world as a whole. Those conscious of the iron grip of depletion might conclude that Russia's national interest would be well served by producing at a low rate so as to make the resource last as long as possible. In commercial terms, it might find advantage in providing its own manufacturers with reliable and even cheap energy, rather than subsidizing its competitors.

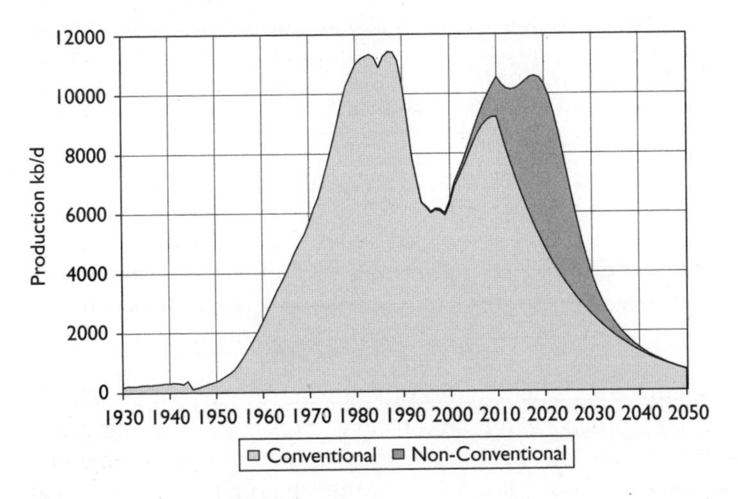

Figure 4.9 Russian production

The Caspian chimera

(See also, Chapter 6.) The Caspian is the oldest oil province in the world, where the Tsars established an oil monopoly even before Col. Drake drilled his famous well in Pennsylvania. The activities have been concentrated on the shores of the Caspian, especially around the early oil center of Baku in Azerbaijan, as the Soviets did have the need to exploit offshore drilling. The area became of great interest to the West at the fall of the Communists, and some wildly exaggerated hopes that it would replace the Middle East were aired, based yet again on the flawed studies of the USGS. The first problem was to decide who owned it: if it was deemed to be a lake, international law required that its resources be jointly exploited by the contiguous countries, a solution favored by Iran and Russia; but if it was deemed a sea, it would be divided up by median lines, as in the North Sea – a solution favored by Kazakhstan. The Western companies, however, moved in without waiting for this little matter to be resolved.

In geological terms, offshore Caspian reserves can be divided into four provinces. In the south lies a deep gas-prone tertiary basin, which has yielded the Shah Deniz Field, operated by BP. To the north is a narrow belt, forming the proto-delta of the Volga, which extends from Baku to Turkmenistan, becoming gas-prone in that direction. The results to date have been disappointing, causing ExxonMobil to withdraw. Future production may not exceed about 10 Gb from known and yet-to-find fields. Next comes a modest Jurassic trend that extends out of Kazakhstan, offering perhaps another 5 Gb. Lastly, in the far north, comes the southern limit of the prolific Pre-Caspian Basin, most of which lies onshore. Interest here was stimulated by the Tengiz Field, found in 1978 by the Soviets, with about 6 Gb of high sulphur oil in a Carboniferous reef at a depth of over 4,000 meters, which is now being developed by Chevron. A huge structure, called Kashagan, was identified in the adjoining waters of the Caspian. Had it been full of oil, it might have justified the exaggerated early claims, but three wells have now been drilled at enormous cost, suggesting that it is made up of several discrete reefs, with a potential in the 10–15 Gb range. At all events the results were sufficiently disappointing to cause BP and Statoil to withdraw from the venture. So, a sanguine estimate suggests that no more than about 30 Gb are likely to be produced from the offshore Caspian, which is equivalent to approximately half the North Sea's resource. This is indeed useful and valuable production, but it is unlikely to have any particular impact on global supply. But it is also approximately equal to the reserves of the US, which may be seen as sufficient justification for its military build-up in the area.

Western Europe

Discovery in the North Sea reached a peak in 1973, with the giant Statfjord Field. Britain exploited its share as fast as possible, partly with advantageous tax provisions, so that production peaked in 1999 and is now set to decline at about 6 percent a year. Norway moved more cautiously, establishing a monolithic state company to take the lion's share, but its production too is now very close to peak, meaning that production in the North Sea as a whole is set to decline to approximately half its present level within ten years (see Figure 4.1, p. 52). It is curious that Norway, having made enormous investments in its state company, should now decide to privatize it so that foreign investors should come to own the priceless national oil and gas patrimony that is set to become infinitely more valuable as world depletion grips.

Europe's imports of oil are set to rise from the current 50 percent to 75 percent by 2010, and to 90 percent by 2025, which will cause a huge drain on its balance of payments as it vies with the US and other countries for access to Middle East oil. Gas imports from Russia, North Africa, and perhaps Central Asia and the Middle East may help reduce the demand for oil until its supply comes to an abrupt, unannounced end, as explained above. Think of poor Ireland, whose demand for electricity has grown with the economic boom. It turned to gas generation, relying on a supply from Scotland, but will soon find itself very much at the end of a line from Siberia, with many energy-hungry countries in between.

Southeast Asia

India, Pakistan, Indo-China, China, and Indonesia have high fertility rates, and find themselves living in a part of the world characterized by convergent plate tectonics that lack rich hydrocarbon source-rocks. China's production is expected to peak around 2003,[2] and all the other countries are long past peak (see Figures 4.10, 4.11, and 4.12). The area has been thoroughly explored, so the chance of a major pleasant surprise is remote indeed. As always, unsubstantiated claims are made for closed areas, including parts of the South China Sea, which are subject to boundary disputes. A certain, though declining, proportion of people in these countries has found out how to live sustainable lives with minimal energy demands, leaving them relatively unaffected by the decline of world oil, but Singapore, Malaysia, South Korea, Taiwan, and Hong Kong are very certainly industrially advanced, energy-intensive economies, and all members of the Association of Southeast Asian Nations (ASEAN) are committed to "conventional" economic growth at the fastest rate possible.

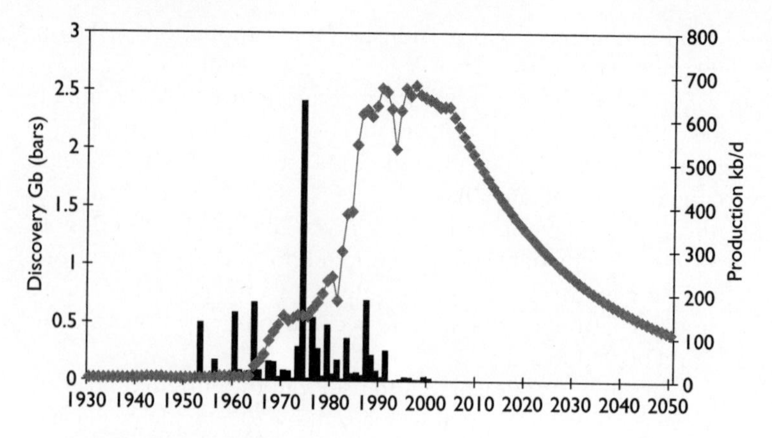

Figure 4.10 Indian production

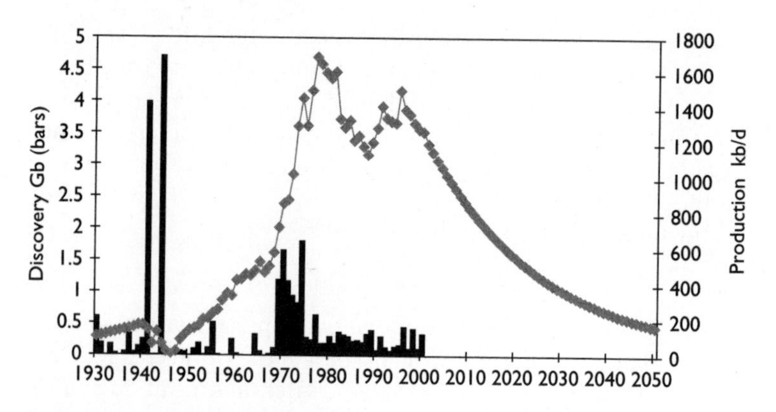

Figure 4.11 Indonesian production

Middle East

Lastly, we turn to review the critical role of the Middle East, which was so uniquely favored with Jurassic source-rocks and effective salt seals to hold the oil within reservoirs. These geological factors combined to make it a concentrated habitat, with most of its oil in a few super-giant fields, found long ago. Exploration has been curtailed since the expropriations, largely because the state companies, lacking the tax inducements available to Western companies, had to fund it out of national budgets, for which there were heavy competing claims. But it is worth noting that the discovery creaming curve (see Figure 4.13) has become very flat, indicating that future discovery will fall far short of past discovery.

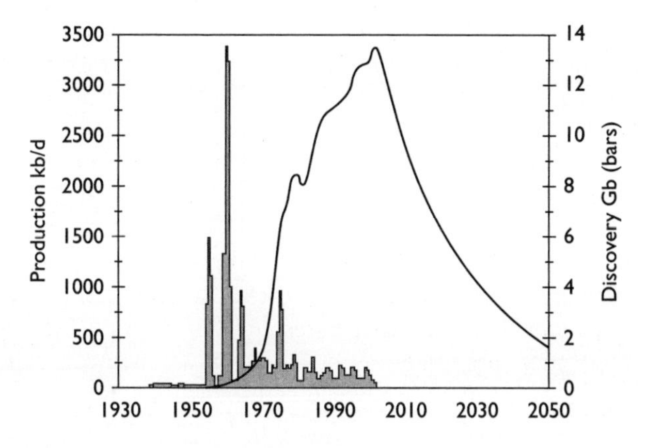

Figure 4.12 Chinese production

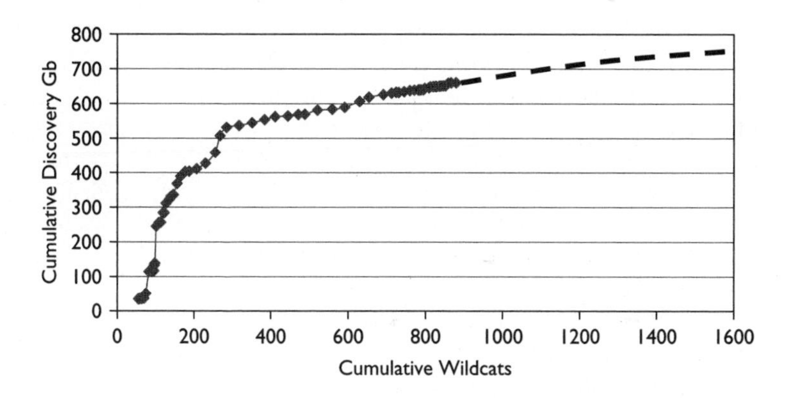

Figure 4.13 Middle East creaming curve

Exactly how much has been discovered is hard to say, because the statistics are exceptionally unreliable, as already mentioned. Kuwait added 50 percent to its reported reserves overnight in 1985, although nothing particular changed in the reservoir. Then in 1988, Abu Dhabi, Iran, Iraq, and later Saudi Arabia responded with enormous increases in retaliation for Venezuela's decision to double its reported reserves by the inclusion of large amounts of long-known heavy oil.

While there are certainly skilled technicians and highly intelligent analysts in the Middle East, the management of the state companies in a highly political environment may be difficult. It is entirely possible that they remain oblivious to what their reserves truly are, possibly still relying on old reports inherited from the private companies before they were expropriated. The OPEC secretariat itself is in no position to question the information furnished to it by its member governments, much as it might be inclined to do so. The assessment is therefore to be taken with reservations. There are growing indications that the reserves are still overstated, although this exaggeration has been recognized by officials in at least some of these countries, and reserves may consequently be revised downwards.

The degree to which the Middle East can continue to exercise its swing role is also uncertain. While the indicated depletion rates are still comparatively low, meaning that in resource terms production can be increased, there are many doubts about how much can be produced in practice. It is commonly claimed that the Middle East has much shut-in capacity, but this is doubtful. Few countries or companies have incentives to drill wells only to shut them in or choke back the production rate, except perhaps briefly, but it is only such wells, which provide spare capacity, that can be brought on at will. Infill drilling, reconfiguring wells and fine-tuning reservoir management all take work, investment, and time to achieve, and the Middle East has to run ever faster to stand still as it desperately tries to offset the natural decline of its aging giant fields. It is reported that Kuwait's wells will soon be producing more water than oil, and the southern end of Ghawar (the world's largest single field) in Saudi Arabia has already gone to water. The demands on the Middle East under the best-case world scenario are illustrated in Figure 4.14. It is far from sure if they are attainable. It is also unlikely that the position will change for the better now that the US has invaded

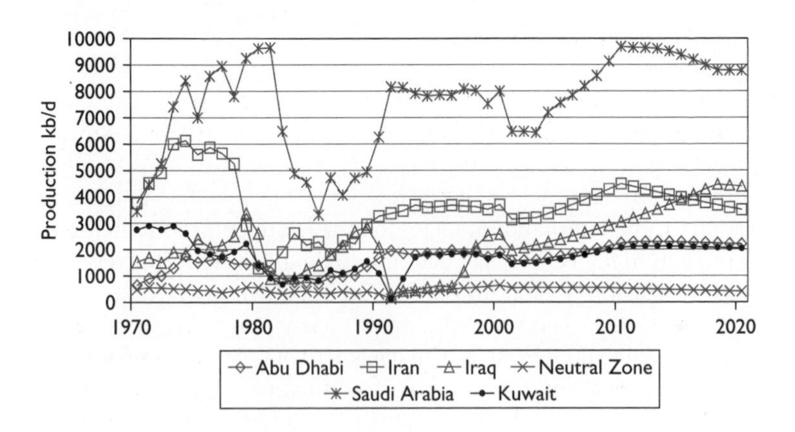

Figure 4.14 Middle East swing production

Iraq, or if it takes the Arabian fields by force. Depletion does not respond to military intervention, and pipelines are easy targets for the vanquished, even using primitive weapons.

CONCLUSION

It is not certain that Darwin got it exactly right with his view of the survival of the fittest. The experience of 500 million years of life on the planet is that species adapted to certain environmental niches and proliferated, only to die out when the environment changed. The limpet, *Lingula*, which prefers a simple life attached to rocks washed by the waves, has survived unchanged since the Cambrian, but more advanced types and forms of life came and went.

Man of human appearance arrived only about 2 million years ago, and the Bronze Age that started him on the path to industrialization began only about 3,000 years ago. He did not give up his flint club because he ran out of flint, but because he found that bronze made a better tool and weapon. The Iron Age followed from small and slow beginnings, but has only dramatically flourished in the last 300 years. At first this age of metals used firewood as fuel for smelting the metal, which in certain countries, such as Denmark and England, led to deforestation before a new fuel was found in the form of coal, lumps of which, known as sea-coal, were at first collected from beaches, before it was mined in shallow pits. Mining itself, as it penetrated below the water table, led to steam-driven machine pumps to drain the surplus water, these pumps being later adapted to provide locomotives for transport.

The fossil-fueled heat engine was developed into the internal combustion engine, driven at first by benzene produced from coal, before turning to petroleum refined from crude oil. This new energy form has transformed the world during the short span of a single century. Cheap and efficient transport opened the world to trade, while the manufacture of consumer goods exploded. The new energy also transformed agriculture, providing the food for a growing population that has expanded six-fold, exactly in parallel with oil production. Oil was in turn followed by gas, increasingly used for electricity generation, which brought power and light to households throughout the world, opening the door to world electronic communication, and eventually the abuse thereof through television, which helped condition the modern consumer mindset and debase human values.

This extraordinary progression was achieved in not much more than 100 years, but it was also accompanied by two world wars, together with related political repression, especially in the Soviet Union, which led to more violent deaths and suffering than the world had ever experienced before.

In the twenty-first century, we face the onset of the natural decline of the premier fuel that made all this possible, and we do so without sight of a substitute energy that comes close to matching the utility, convenience and low cost of oil and gas. It remains to be seen if we will be the only species in over 500 million years of recorded history to evolve backward, from complexity to simplicity. Don't hold your breath, but there is a little time left to adjust, as we have about as much oil left as we have used so far. Our challenge is to maintain demand in pace with or below the depletion rate. The first step in that direction is to determine what the depletion rate is, and to inform ourselves better about the resources with which nature has endowed us.

NOTES

1. API refers to the American Petroleum Institute scale, measured in degrees of density.
2. EIA Crude figures to 2006 show that China's production did not in fact peak in 2003. Discoveries in 2007, although the largest in 40 years, were predicted to boost China's known oil reserves by 20 percent, to yield annually somewhere around 180,000 to 200,000 barrels a day, equivalent only to China's growth in foreign oil imports that year. http://news.mongabay.com/2007/0508-china.html.

5
Coal Resources of the World

Seppo A. Korpela

COAL RESOURCES

Coal was formed in the Carboniferous era between 360 and 286 million years ago from trees and other plant material located in low relief sites near ancient seas. There the plants near swamps and peat bogs became buried under water during periods of rising sea level. In the anoxic bottom layers sub-aqueous bacteria working on the plants released part of the oxygen, nitrogen, and hydrogen in the organic molecules, leaving mostly carbon. In the deeper layers heat and pressure then completed the process of coalification and turned peat first into coal that is soft and brown and then into hard and black coal.

The deeper the burial, the higher the carbon to hydrogen ratio of coal, and in *anthracite,* a coal rank both oldest and hardest, hydrogen is nearly absent. Only 1 percent of the reserves of the world are anthracite, and it never was abundant. It was mined for home heating, and lack of volatiles caused it to burn cleanly without a visible flame.

During the Upper Carboniferous, or the Pennsylvanian period, large swamps existed throughout the eastern parts of the North American continent. Much of the plant matter in them turned into hard coal, with anthracite in eastern Pennsylvania and bituminous coal in the Appalachian field extending from Pennsylvania to Alabama.

Bituminous coal is also a hard coal and the best seams have heating values better than anthracite's, owing mainly to the larger fraction of inert material in the latter. This contributes little to the heating value and upon burning becomes ash.

The world's hard coal is used mostly for electricity generation and for metallurgical processing of iron. Another important use is in the manufacture of cement. The coal burned in steam power plant boilers is called steam coal to distinguish it from coal diverted to making coke for steel plants. It is bituminous coal which has powered the world since the beginning of the industrial revolution.

The next lower in coal rank, and thus having a lower heating value, is sub-bituminous coal. The deficit in heating value is proportional to the amount of moisture in the coal, for the water present just turns into vapor when coal is burned. In the United States sub-bituminous coal is found in large quantities in the state of Wyoming and today the share of this lower rank coal in the production mix is nearly the same as that of bituminous coal. For the entire world 53 percent of reserves are hard coal.

Lignite, the youngest, softest, and lowest quality coal, has a heating value only half that of the best bituminous coal seams. Large deposits are found in the state of Montana, in the United States. In lignite, plant matter is still visible and the composition of this coal is close to that of peat, the heating value of which is about the same as that of dried wood.

Unlike petroleum, which migrates from its source to reservoirs, coal, once formed, remains in the layered strata in seams reaching 35 meters thick.

Most of the world's coal occurs in regions north of 30th parallel. Only in Australia and South Africa are there substantial deposits of coal in the Southern Hemisphere. In every coal region there are large horizontal variations in the quality of the coal beds, as might be expected based on an uneven burial history. Vertical variation is pronounced in mountainous regions in which the folding of the earth's crust in the geological past has led to this unevenness. Steeply angled seams are difficult to mine, as are those that are too thin.

In reporting the energy density of coal the notation MJ/kg means million joules of energy per kilogram of coal. An equivalent expression is GJ per tonne, in which GJ refers to gigajoules, or billion[1] joules of energy. Energy in various ranks of coal is shown in Table 5.1.

Table 5.1 Energy content of different ranks of coal

Coal Rank	*Range*	*Average*
Anthracite	30 MJ/Kg	30 MJ/kg
Bituminous	18.8–29.3 MJ/kg	24 MJ/kg
Sub-bituminous	8.3–25 MJ/kg	17 MJ/kg
Lignite	5.5–14.3 MJ/kg	10 MJ/kg

Source: *Coal: Resources and Future Production*, Energy Watch Group.[2]

The *BP Statistical Review of World Energy* assigns 28 MJ/kg for hard coal and 14 MJ/kg for soft coal. Thus soft coals have one half of the heating value of the coals from the best seams. This makes the international trade in soft coals unprofitable, and even domestic transport by train is expensive. In the US, sub-bituminous coal is shipped from Wyoming to the eastern parts of the country, where it is blended with Appalachian coal of higher sulphur content in sufficient quantities to keep the sulphur emissions below legal limits. The transportation cost for utilities in the United States in 1979 was 23.4 percent of the cost of coal; by 1997 this had increased to 35.9 percent. The transportation cost of the lowest quality coal can be as high as 70 percent of the price. For this reason lignite is often used in power plants at the mouths of mines.[3]

COAL RESERVES, PRODUCTION, AND CONSUMPTION OF COAL IN THE WORLD

Over 90 percent of world coal reserves exist in the 13 countries listed in Table 5.2, and the top six have slightly over 80 percent of the reserves.[4] The world's total reserves are seen to be 478.8 Gt, or gigatonnes, of hard coal and 430.3 Gt of soft coal, with a total of 909 Gt. A slightly lower number of 857 Gt is reported by the World Energy Council.[5] Of the reserves, one-quarter is in the United States.

Table 5.2 Coal reserves of major producing countries in billions of metric tons (Mt)

	Country	*Hard, Gt*	*Soft, Gt*	*Percent*	*Energy, MJ/kg*
1	USA	111.3	135.3	27.1	19.6
2	Russia	49.9	107.9	17.3	17.7
3	China	62.2	52.3	12.6	18.3
4	India	90.1	2.4	10.2	26.6
5	Australia	38.6	39.9	8.6	20.1
6	South Africa	48.8	0	5.5	27.0
7	Ukraine	16.3	17.9	3.8	19.3
8	Kazakhstan	28.2	3.1	3.4	25.6
9	Poland	14.0	0	1.5	27.0
10	Colombia	6.2	0.4	0.7	26.2
11	Canada	3.5	3.1	0.7	20.8
12	Germany	0.2	6.6	0.7	13.9
13	Indonesia	0.7	4.2	0.5	15.5
	World	478.8	430.3	100	20.6

Source: *BP Statistical Review of World Energy*, 2007.

The earliest estimate of the world's ultimate recoverable reserve was reported to be 7,400 Gt at the 12th International Geological Congress held in Toronto in 1913, of which 3,840 Gt was in the United States. The amount for the world was reduced to 3,180 by Hendricks in 1939 and to 2,419 Gt in 1960 by Averitt.[6] About 99 Gt had been consumed by year 1960, leaving 2,320 Gt. At end of 1971, in the 1974 Survey of World Energy Resources, the estimate of recoverable coal was increased to 3,000 Gt, and by 1971 cumulative production stood at 130 Gt.

Today the recoverable reserves are 909 Gt, a large drop from 3,000 Gt in 1971. From the tally of cumulative production the drop cannot be attributed to production in the intervening years. Better information on the angle of dip of coal seams, their thickness, and the economics of mining poor seams are contributing factors. The locations of the world's prolific coalfields are well known and finding them does not depend on exploration success, as in the hunt for oilfields. Thus the natural tendency in coal reassessments ought to be downward revisions.

For the US in 1961 the minable coal amounted to 753 Gt, in 1971 it was 510 Gt, and today it stands at 257 Gt. Again, the downgrading of the reserves, rather than production during the intervening years, is the cause for the diminishing reserves. In addition, the *BP Statistical Survey* shows that since 1998, and perhaps longer, the reported reserves in the US and China have not changed at all. During this short interval in the US 10.5 Gt of coal has been produced and in China the production amounted to 15.5 Gt. The latter amount is nearly 10 percent of China's reserves. India's reserves have increased by 24 percent, and for the entire world the tally shows a decrease of 8 percent.

The production and consumption rates, together with trade, for the year 2006 are listed in Table 5.3 for the main coal producers of the world.

China's production rate stands out as it is by far the largest producer. Its reserve to production ratio has dropped to 74 and it is one of the lowest on the list. Recently China was still able to export some of its production to Japan and South Korea, but in late 2006 the government imposed a 5 percent export tariff on coal, which reduced exports to nought. The voracious demand for energy in China has now not only erased its surplus, but has made China a net importer of coal. The imported coal comes from Australia and Indonesia.[7]

Table 5.3 Coal production, consumption, trade, and reserve to production (R/P) ratios for major countries for the year 2006 in millions of metric tons (Mt)

	Country	*Production Mt*	*Consumption Mt*	*Exports Mt*	*R/P Years*
1	US	1,053.6	1,004.4	49.2	234
2	Russia	309.2	240.7	68.5	508
3	India	447.3	507.0	–59.7	207
4	China	2,380.0	2,338.7	41.3	74
5	Australia	373.8	94.0	279.8	210
6	South Africa	256.9	166.4	90.5	190
7	Ukraine	80.5	76.3	4.2	387
8	Kazakhstan	96.3	58.1	38.2	325
9	Poland	156.1	136.1	20.0	90
10	Colombia	65.6	3.7	61.9	101
11	Canada	62.9	68.2	–5.3	110
12	Germany	197.2	323.0	–125.8	34
13	Indonesia	195.0	45.1	149.9	25
	World	6,195.1	6,216.0	n/a	147

Source: *BP Statistical Review ofWorld Energy*, 2007.

In India, despite its large coal reserve, consumption already exceeds production, making India a net importer. Much of the imported coal comes from South Africa.

Russia has the largest reserve when compared to its domestic use. It will remain self-sufficient for a long time. Russia is also a large exporter of oil, and its natural gas reserves are the largest in the world. These have made post-Soviet Russia an important energy supplier to Europe, where Germany is the largest coal user, and despite a reserve to production ratio of 34 years, it imports over one-third of its coal. Its hard coal reserves are quite meager, even if it still has a considerable amount of soft coal. Poland's excess coal production is exported to Germany, Central Europe, and Scandinavia. The United Kingdom's coal production is 19 Mt and consumption stands at 72 Mt. Added to the UK's energy dilemma is the passing of its oil production peak in 1998, with the result that today the UK is a net importer of oil. Similarly, its natural gas reserves from the North Sea are dwindling rapidly.

The world's largest coal importer is Japan with 219 Mt in 2006, followed by South Korea with 101 Mt, and Taiwan, importing 73 Mt. Since Australia, as the world's largest coal exporter, provided 280 Mt to world markets in 2006 from its production of 374 Mt, the rest of the demand in Japan, South Korea, and Taiwan is met from countries such as Indonesia, the other major exporter in Asia. Indonesia's reserve to production ratio is the lowest of the major producers and it will soon cease to be an exporter of coal. Indonesia still has natural gas to export, but its oil production is dwindling. About 700 Mt of hard coal is traded internationally and over 90 percent of this is by sea.

COAL PRODUCTION IN THE US

An estimate of future US coal production is shown in Figure 5.1, which is based on Hubbertian analysis using the logistic equation, with early data from Putnam[8] and later data from the EIA.[9] Hubbert's model is located in Chapter 3, "Prediction of World Peak Oil Production" in this volume. US coal production trajectory calculation was recently carried out by Rutledge,[10] also using a model based on the logistic equation. The analysis here gives an estimate of 238 Gt for US ultimate

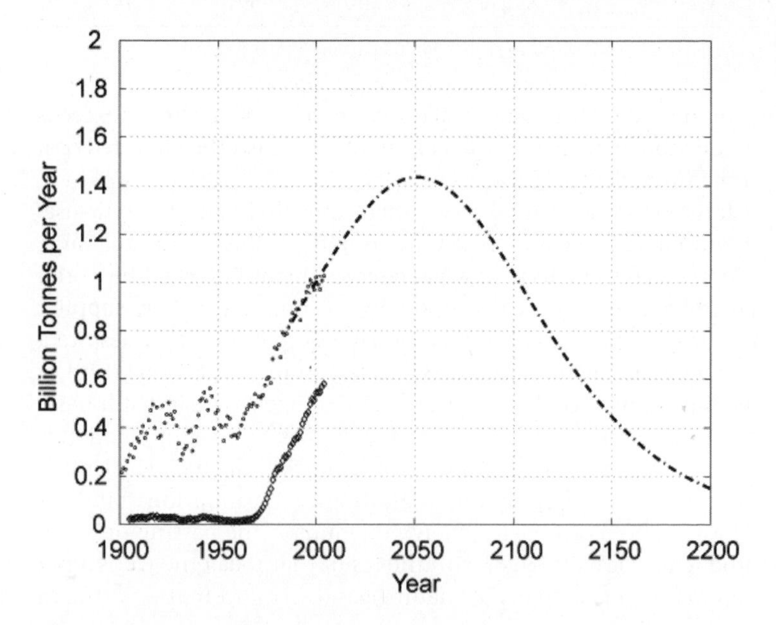

Figure 5.1 US coal production trajectory

Circles indicate the actual production and dashed line is obtained from a model based on the logistic equation. The lower diamond shapes are production of coal west of Mississippi River.

Sources: 1900–49: Palmer C. Putnam, *Energy in the Future*, Van Nostrund Co., New York, 1953; 1949–2006: EIA-DOE.

recoverable coal reserve. Since the cumulative production stood at 58.6 Gt at the end of 2005 and with the present reserves at 247 Gt, the ultimate recoverable reserve should be about 306 Gt. Thus the difference in the estimate of the ultimate recovery and that based on present reserves and cumulative production is quite large, but if the trend seen in other countries in the downgrading of their coal reserves also takes place in the United States, then the estimate of 238 Gt might be close to the actual.

The peak production at 1.42 Gt is estimated to take place in 2051. It is known that a model based on the logistic equation is unreliable if it is applied to predicting a peak that lies far in the future, and 43 years is a rather long time. How much the estimates obtained from the logistic equation model change as more data become available is today unknown. But it is certain that as years go by, the model estimates improve when each year's data become part of the record. Thus the mid-century mark is a good tentative estimate for the peaking of US coal production. It is certainly better than any guess based on the 234-year reserve to production ratio.

US coal reserves have not been reassessed since the early 1970s. The data shown in Figure 5.1 indicate that the eastern bituminous coal production is no longer growing, and the present growth is entirely based on the increase in the western soft coal. Thus coal quality is decreasing. Even if – with the world's largest reserves – the United States can be expected to remain self-sufficient in coal for some time, it is as well to remember that the US became a net importer of oil already in 1949 despite being able to increase oil production until 1970. Were coal consumption to follow a similar pattern the question is, from where would this imported coal come? Already in 2005 the US imported 28 Mt of hard coal, and besides Colombia there are few places left from which to acquire imports.

COAL PRODUCTION IN CHINA

It was noted above that China is the world's largest producer of coal by a wide margin, for it produces more than double the US figure. Its reserve base is 114.5 Gt according to the 2007 *BP Statistical Review*, although the Chinese Ministry of Land and Natural Resources claims 186.6 Gt.[11]

Early coal production in China concentrated in Manchuria and along the eastern seaboard with production reaching 35 Mt in 1936. It rose to 60 Mt in 1943, but then sank to 15 Mt by 1946. By 1962 it had passed its prewar high, and from there on it has continued its astonishing growth to 2,380 Mt a year today. Much of the mined coal comes also from the western province of Xinjiang as well as from Inner Mongolia and Shanxi, the latter being also a good oil-producing region.

China's coal mines consist of those operated by state-run coal companies, which are administered by provincial governments. There are also local state mines, which may be administered by county governments. About 17,000 mines still operate at the village level.[12] As late as 1958, armies of peasants worked the mines with picks and shovels, with coal hoisted by ropes wound to millstones which were turned by bullocks.[13]

In the late 1990s the government of China began to close the small mines in order to modernize the mining industry. This shows up in the coal production statistics as a decreasing coal production from 1997 to 2001. After this, coal production has increased very rapidly at the compounded yearly rate of nearly 12 percent.

Owing to the recent spurt in coal production, it is difficult to apply a logistic model to China's coal production. To obtain a tentative estimate, the intrinsic growth rate was fixed at 7 percent and the ultimate production set at 190 Gt. This yields the production profile shown in Figure 5.2. The maximum production rate takes place in the year 2021 at the yearly rate of about 3.3 Gt. By using a different method, Tao and Li propose that peak production takes place later in 2029 at peak production rate of 3.8 Gt per year. Their ultimate recoverable reserve is 223 Gt.

The results are in reasonable agreement, and the later date is made possible by assuming the larger reserve base.

Much coal is needed in China's large coastal cities and it is advantageous to import it, rather than haul it over long distances from its western province. The *International Energy Outlook* for 2007 expects China and India to account for 72 percent of the increase in world demand for coal from here until 2030.[14]

WORLD COAL PRODUCTION

Owing to the rapid increase in world coal production over the last ten years, Hubbert's model does not give reliable results for predicting the future coal production in the world. The recent data are plotted in logarithmic form in Figure 5.3. It shows that the world production from 1965 to 2006 has grown by 1.8 percent.

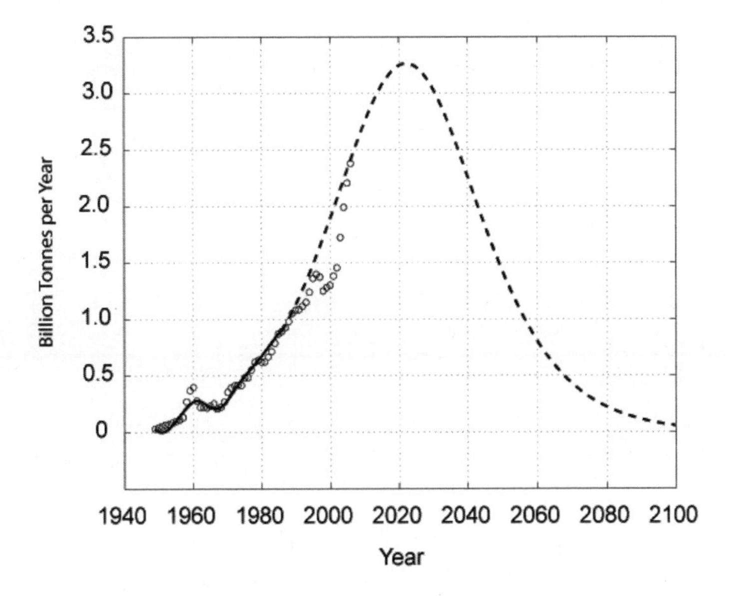

Figure 5.2 China's past and future coal production

Circles indicate the actual production and the dashed line is obtained from a model based on the logistic equation with intrinsic growth rate of 7 percent and ultimate recoverable reserves of 190 Gt.

Source: Z. Tao and M. Li, "What is the Limit of Chinese Coal Supplier – A STELLA Model of Hubbert's Peak," *Energy Policy*, vol. 35 (2007), pp. 3145–54.

With China's coal production reaching its peak in about dozen years and the US peaking by the end of the first half of this century, it is reasonable to expect that the world's coal production will also peak before 2050. The Energy Watch Group, in a report issued recently, reached this conclusion and placed the likely peak as early as 2025. Their report is based on the recent reassessment of the coal reserves by BGR, Germany's Federal Institute for Geosciences and Natural Resources. The work of BGR was highlighted at the ASPO-5 conference by Gerling, who showed the BGR's downgrading of coalfields in the Russian coal regions of Donets and Kuznetsk, as well as those in the Upper Silesian region of Poland and Ruhr in Germany.[15]

Reserve estimates over the years have been reduced by large amounts. In 1924 it was thought that the United Kingdom's coal reserves would last for 350 years based on the simple calculation of reserve to production ratio, for the production rate was quite low.[16] The reserve to production ratio has now dropped to nearly

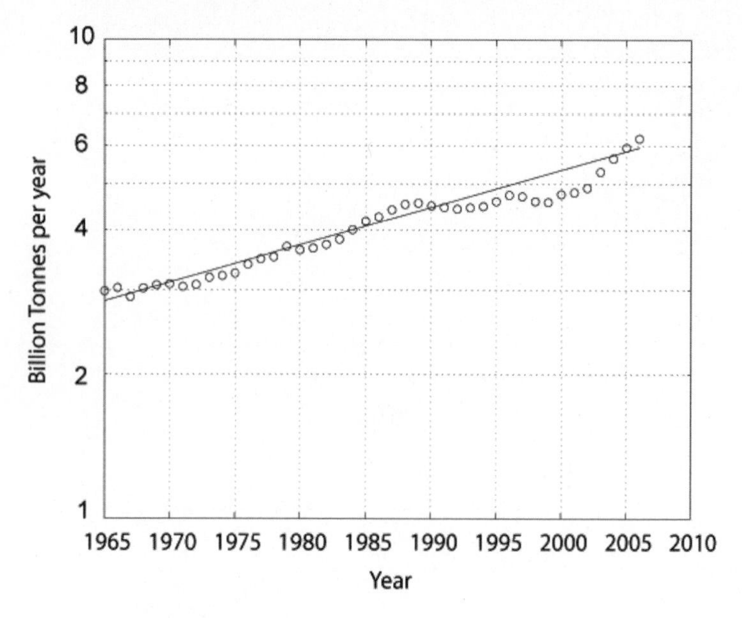

Figure 5.3 The growth of world coal production since 1965

Growth is plotted to show that it has increased 1.8 percent since 1965. Since 2002 the growth has accelerated.

Source: *BP Statistical Review of World Energy*, various editions.

zero as very few mines are left in operation. A similarly large decrease in reserve to production ratio over the last 100 years has brought Germany's coal mining to a standstill, and even Poland's reserves were reduced by half by BGR's latest assessment. The Energy Watch Group points out that only India has increased its reserve base for coal and that South Africa's reserve to production ratio follows the natural decline as coal is produced each year. For other countries the reserve to production ratio falls as a result of downgrading the reserves. This has also been true for the US coal reserve, which has decreased substantially over the last century.

The world reserve to production ratios are shown in Table 5.4 for the last few years. These show a 32 percent reduction over the last eight years. This rapid reduction must moderate as time passes.

Space does not permit an extensive discussion of the future of industrial civilization. But it should not tax the imagination of the reader to see that it is in peril. Humanity has landed itself in a trap

Table 5.4 Reserve to production (R/P) decrease for world coal 1998–2006

Year	*1998*	*1999*	*2000*	*2001*	*2002*	*2003*	*2004*	*2005*	*2006*
World R/P	218	230	227	216	204	192	164	155	147

Source: *BP Statistical Review ofWorld Energy*, various editions.

from which it cannot escape. The built infrastructure demands huge energy and material flows and, as these flows diminish, so does the complexity of modern societies. How to manage the transition is the question that needs to guide our thinking from now on.

NOTES

1. US billion, or 1,000,000,000.
2. All internet resources accessed December 12, 2007. Energy Watch Group, *Coal: Resources and Future Production*, Report EWG-Series No. 1, March 2007, www.energywatchgroup.org/fileadmin/global/pdf/EWG-Coalreport_10_07_2007.pdf.
3. *The Coal Resource: A Comprehensive Overview of Coal*, Coal Institute, www.worldcoal.org/assets_cm/files/PDF/thecoalresource.pdf.
4. "Coal," *BP Statistical Review of World Energy* 2007, pp. 32–5, www.bp.com/productlanding.do?categoryId=6848&contentId=7033471.
5. World Energy Council, *Survey of Energy Resources 2007*, www.worldenergy.org/publications/survey_of_energy_resources_2007/default.asp.
6. M.K. Hubbert, *Energy Resources*, National Academy of Sciences – National Research Council, 1962.
7. Yangtze Yan, ed., "China's Coal Exports Down, While Imports Up as Export Tariff Bites," *Window on China*, October 18, 2007, http://news.xinhuanet.com/english/2007-10/18/content_6901616.htm, accessed December 14, 2007.
8. P.C. Putnam, *Energy in the Future*, D.Van Nostrand Co. Inc., Princeton, New Jersey, 1953.
9. Energy Information Administration (EIA), US Department of Energy (DOE), *International Energy Outlook*, Chapter 5, *Coal*, 2007, www.eia.doe.gov/oiaf/ieo/pdf/0484(2007).pdf.
10. D. Rutledge, *Hubbert's Peak, the Coal Question and Climate Change*, video and PowerPoint, http://rutledge.caltech.edu/.
11. Z. Tao and M. Li, "What is the Limit of Chinese Coal Supplies – A STELLA Model of Hubbert's Peak," *Energy Policy*, vol. 35 (2007), pp. 3145–54.
12. "China to Close 5000 Coal Mines to Improve Safety," AFP, September 4, 2006, *Terra Daily*, www.terradaily.com/reports/China_To_Close_5000_Coal_Mines_To_Improve_Safety_999.html.
13. J. Dwyer, "The Coal Industry in Mainland China since 1949," *Geographical Journal*, vol. 129 (September 1963), pp. 329–38.
14. EIA-DOE, *International Energy Outlook* 2007.

15. Peter J. Gerling, T. Theilemann, and S. Schmidt, "Coal – The Prime Energy Fuel of the Future?" Presentation at the Association for the Study of Peak Oil Conference, ASPO-5, Pisa, Italy, July 2005; Energy Watch Group, "Coal: Resources and Future Production."
16. Putnam, *Energy in the Future*; E. Ayres and C.A. Scarlott, *Energy Sources – The Wealth of the World*, McGraw-Hill, New York, 1952.

Part II: Geopolitics

GEODESTINIES AND GEODETERMINISM

"History is subject to geology"[1] wrote Will and Ariel Durant in 1970, and petrogeologist Walter Youngquist cited them in 1997 in his book, *GeoDestinies*. Youngquist concluded his book by writing: "As it always has been and always will be, the materials of the Earth: soil, water, metals, and energy supplies, will be the base for civilization and control its destiny."[2]

The chapters in this part explore the geopolitics of petrocarbon supply and demand which underpin our civilization and control its destiny. The Ancient Greeks believed that human destiny was entirely in the hands of the gods. From this came the concept of "irony" whereby individuals struggled heroically in the pathetic belief that they could affect their own destinies. Plays about this struggle could inspire laughter or tears in the audience. Psychoanalytically it was all about individuals who found, to their horror, that they were behaving in ways they had never endorsed or intended which pushed them inexorably to a certain end.

Humans have always argued about the degree of control they could have over their fate in various manifestations of determinism. This is one basis of divisions between peak oil groups where, for instance, some argue that "we" can change our collective behavior and "powerdown" gracefully, but others maintain that our behavior is "hard-wired" and that all creatures are programmed to multiply, overshoot, and crash.

Between these perceived oppositions of "free will" and universal determinism, are forms of limited determinism. For instance, Descartes believed that thought was independent of physical constraints and thus that reason had the power to identify causality and thence to avoid certain outcomes. Medieval Catholicism was more Greek with its determinism, endorsing original sin and the concept that people were born to set pathways from which they might only be promoted or escape after death. Protestantism took the view that humans could affect their destiny to a large degree by cultivating certain habits, such as industriousness, frugality, and faithfulness, which might then find their reward well before death in well-deserved earthly riches and status. All this was, of course, determined by overarching divine laws.

Capitalism took its cue from such Protestantism, but there is also a theory of North American history called "manifest destiny," whereby postcolonial North Americans embrace a well-defined pathway to national triumph, using tools they were "meant" to find along the way, such as minerals and science. This view sees capitalism as founded on natural laws with North Americans destined to do very well out of those laws.

Marxism was very deterministic in that it identified an imposed system and then worked logically to overcome this in order to release the rightful entitlements of the proletariat. Nonetheless, Marxism still adhered to the idea that every man or woman is born to earthly wealth which humans will find tools to extract. In both Marxism and capitalism, Destiny is renamed Progress.

When I was thinking of a way to reintroduce this part, taking account of the time which had passed between the first and the second edition, I focused on "Battle of the Titans," by the late Mark Jones. Mark, once an adviser to LUKoil (Russia), wrote the chapter in 2002 for the first edition of *The Final Energy Crisis*. His chapter identifies economic and political pressures which make and break geopolitical hegemonies. Mark thought the US had a window of only a few years to try to gain control of foreign oil reserves before it would lose its economic and military advantages to China. As editor I wondered how to update this article or follow it up, without Mark. Then I thought about the surprising new geopolitical hub at the mouth of the exotic and threatened Amazon, where an unexpected realignment of factions is being led by Venezuela's almost mythologized President, Hugo Chávez. I have described this situation in a chapter later in this part about Chávez and Latin American Oil.

Going into this Latin American situation a little more deeply, I realized that here again the notion of destiny and free will were being played out, in rather gothic representations of free market capitalism vs. Chávez Christian Socialism.

Now, what could I possibly mean by this? Well, rather than endorsing the continued ironic struggle of Venezuela's citizens by allowing the Market, like the Greek gods, to determine their socio-economic destiny, President Chávez has apparently imposed human rules on the division of petro-wealth and has furthermore encouraged a philosophy of conservation and consolidation whereby Venezuelans, and Latin America itself, will join with other nations which are downtrodden, but relatively richly endowed with petroleum, to bring about a new world order antithetical to the G7 corporate world order.[3]

However, according to Anglophone sources, like *The Economist* or the *New York Times*,[4] everything good that has come about in Venezuela since Chávez was due to earlier economic reformist regimes and the prediction is that, with Chávez, Venezuela and Latin America are foolishly and willfully selecting oil economics options which will guarantee them ignominious ends. Latin America is defying the gods!

GEODETERMINISM

Obviously oil reserves control our destiny, but the question is, to what degree? Are we indeed hardwired to overrun the earth as if it were a Petri-dish and we were cancer cells? Controlling our destiny depends on how well we are able to control demand, which means controlling our collective numbers and consumption. It is not just the oil-importing countries which find it hard to restrain their consumption; the producers and exporters are just as hooked on the money that comes from sales. For this reason market-based economies are a very flawed approach to "powering down." This is unfortunate because most oil supply is controlled by private companies over which governments have very little control.[5]

> The Soviets were very efficient explorers, as they were able to approach their task in a scientific manner, being able to drill holes to gather critical information, whereas their Western counterparts had to pretend that every borehole had a good chance of finding oil. (Colin Campbell, "The Caspian Chimera," Chapter 6 in this part)

As Campbell implies, private industry cannot afford to be careful and conservationist in its exploration and exploitation of geological resources because it has to make a profit. In the incredibly expensive industry of oil exploration, results have to be obtained really fast to satisfy investors. Supplies cannot be delivered in a measured way according to need and keeping finite reserves in mind because the output must pay for the costs, not only of exploration, but the costs of profit, which include massive salary packages, major marketing, and high shareholder returns on high-risk investment.

Those private profits have to be big on a similar scale to the large private ventures. Profits must be seen to grow constantly, even though risks are rising and real returns are declining worldwide along with accessible oil.

As Andrew McKillop wrote in the first edition introduction to this part, "gas, like oil, is a traded commodity. Any oversupply and the projected or actual price spikes turn into price crashes. Investments running at tens or hundreds of billions of dollars or euros simply cannot be financed in the face of such uncertainty."

State-based activities offer much greater ability to control such costs. When resources are managed as necessities there is, furthermore, no obligation or commercial gain in using them up as fast as possible. A state may avoid all the costs of commercialization which dog private enterprise every step of the way.

Yet learned values lead many people to mistake the ideology of profit for a law of nature. To read some economic rationalist newspapers you would think that state-based enterprises and the politicians who espouse them were all doomed by biological evolution. It is, however, always at some stage an economic and political choice whether land, soil, material and mineral resources are treated as commodities or as necessities, just as it is a political choice whether wealth and amenity are shared among citizens or mostly accumulated within a restricted class.

Beliefs, however, are themselves social constraints, which collectively form systems that push societies along certain paths. Arguably profit-oriented societies hardwire profit-making paths in the face of massive environmental and social costs and create a tragic destiny in the same way that noise, lights, and unpredictable payouts on a poker machine hardwire a gambler to ignore the destruction of his personal life. Without that kind of reinforcement, however, more sensible systemic feedback is possible. James Q. Wilson devised a political theory which supports this kind of explanation.[6] I have mentioned in an earlier chapter in this volume[7] how this was the case with France and Western continental Europe after the first oil shock. It may also become the case for Latin America. In the world a struggle seems to be emerging between the global free market gods and human societies. The market can only rule while there are plenty of resources. Perhaps a Chávez-led international political movement is a sign that the global free market is losing its hegemony over human destiny and a signal of the coming or actual increasing scarcity of resources.

Two other chapters in this part by Andrew McKillop are about Africa and China – at extreme opposites of the growth machine. As McKillop writes, "the oil-hungry eyes of OECD national leaders will remain riveted on the Dark Continent for one reason: because it is so dark," meaning that foreign aid and development for Africa somehow manage to keep Africa without electricity and Africans without hope, thus ensuring that there will be more oil to export to supply continuous growth elsewhere – in China, for instance. And will this terrible arrangement persist for long enough for the world to boil in China's exhaust gases? Can Al Gore and the Intergovernmental Panel on Climate Change stop this? Can a Chávez-led movement stop this? Or are the gods really crazy?

NOTES

1. Will Durant and Ariel Durant, *The Lessons of History,* Simon and Schuster, New York, 1970, cited in W. Youngquist, *GeoDestinies,* National Book Company, Portland, Oregon, 1997, p. xv.
2. Youngquist, *GeoDestinies*, p. 474.
3. New relationships are forming between Latin American and Ex-Soviet Union and other third world parties, involving national facilities and companies like Mexico's Petróleos Mexicanos, Russia's state-owned export pipeline, Transneft, and oil exploration company, Rosneft (formed by nationalizing Yukos); China's CNPC and India's ONGC which have engaged to work together in Africa, Central Asia, and Latin America. Amitav Ranjan, "Let's Shop for Oil, Gas Together," *Indian Express*, August 24, 2005, www.indianexpress.com/res/web/pIe/full_story.php?content_id=76875, accessed October 5, 2007.
4. "Chávez vs Bush," *The Weekend Economist* "Quaerere Verum," March 9, 2007, http://weekendeconomist.blogspot.com/2007/03/47-chavez-vs-bush.html, accessed October 7, 2007; James Neophytou, "Economist's Obsession with Chávez," *Latin American News Review*, May 5, 2007, http://lanr.blogspot.com/2007/05/economist-s-obsession-with-chavez.html, accessed July 20, 2007; Simon Romero, "Moving Clocks Ahead, Reaching Back in Time," *New York Times*, August 26, 2007, www.nytimes.com/2007/08/26/weekinreview/26romero.html?_r=1&oref=slogin, accessed October 7, 2007.
5. EIA, "Non-OPEC Fact Sheet," Country Analysis Briefs, www.eia.doe.gov/emeu/cabs/nonopec.html, accessed April 10, 2007.
6. Political scientist James Q. Wilson devised a method that classified four types of politics depending on whether the benefits and costs of policies were concentrated or diffuse. Narrowly focused benefits mean that those benefiting from something are usually conscious of this and are able to recognize each other and organize to keep those benefits flowing. Where costs are diffuse, falling upon a disparate population at many different points in many different ways, they are difficult to identify and there are no obvious political rallying points for the public to organize a protest around. J.Q. Wilson, ed., *The Politics of Regulation*, Harper, New York, 1980.
7. Chapter 2, "101 Views from Hubbert's Peak."

6
The Caspian Chimera

Colin J. Campbell

The Caspian is one of the most ancient oil provinces of the world. The Zoroastrians of antiquity worshipped the eternal flames of Baku, which were smoldering hydrocarbon source rocks and gas seepages. F.N. Semyenov drilled a well there in 1840, operating under a concession granted by the Tsar of Russia, eleven years before the self-styled Colonel Drake drilled his well at Titusville, Pennsylvania, which is commonly taken to mark the start of the modern oil industry.

Geographically, it is a saltwater inland sea or lake covering about 375,000 square kilometers, bordered by the Elburz Mountains of Iran to the south and the Caucasus to the northwest. The Volga River flows into it from the north, forming a large delta near Astrakhan, but evaporation is sufficient to counter the influx, leaving it some 30 meters below world sea level. It is flanked to the north by Russia itself, followed clockwise by Kazakhstan, Turkmenistan, Iran, and Azerbaijan. The three "-stans" gained independence following the fall of the Soviets in 1991. Dagestan and Chechnya, which are still Moslem provinces of Russia on the shores of the Caspian, are still seeking their independence, in a vicious campaign attended by many acts of terror. Under international law, ownership of the offshore mineral rights depends on whether it is deemed a lake or a sea. In the case of a lake, they belong jointly to the contiguous countries, whereas in the case of a sea they are divided up by median lines. The matter, which is no small issue, has yet to be fully resolved, but it seems in practice to be moving in the direction of the latter formula. It is worth noting here that Tehran, the capital of Iran, lies only 100 kilometers from the Caspian shore, so its role in the future of the region cannot be ignored.

In geological terms, it is made up of several diverse provinces. To the south there lies a deep Tertiary basin in the fore-deep of the Elburz Mountains. It is followed to the north by the proto-delta of the Volga that runs across the Caspian as a fairly narrow belt from Azerbaijan to Turkmenistan. That gives way to a Mesozoic basin, running out of Kazakhstan, which in turn adjoins the southern part of a large Paleozoic basin, known as the pre-Caspian basin, whose axis lies to the north of the Caspian.

Early oil activities were concentrated on the Aspheron Peninsula of Azerbaijan, around the town of Baku on the proto-delta of the Volga. Oil and gas, generated in lower Tertiary deltaic sediments, has migrated upwards, mainly along fault-planes, to accumulate in a thick sequence of Miocene and Pliocene sandstone reservoirs at fairly shallow depths. A peculiar feature is the so-called mud volcano, in which gas seepages carry mud to the surface giving volcano-like features, several hundred meters high, which occasionally ignite and explode. Extensions of this same geological province extend northwards into Chechnya, where many people still make a living refining oil from shallow wells and seepages in primitive, dangerous, and very polluting home-made stills.

Baku was one of the great world oil centers during the late nineteenth century. The Nobel brothers of Sweden held a dominant stake, later joined by the Shell Oil and Rothschild interests that financed a pipeline to the Black Sea. No less a figure than Joseph Stalin had his early experiences in Baku as a workers' leader facing the appalling operating conditions of the early oilfields, a ready breeding ground for revolutionary ideas.

The Soviets were very efficient explorers, as they were able to approach their task in a scientific manner, being able to drill holes to gather critical information, whereas their Western counterparts had to pretend that every borehole had a good chance of finding oil. In the years following World War II, they brought in the major producing provinces of the Union, finding most of the giant fields within them. Baku was by now a mature province of secondary importance, although work continued to develop secondary prospects and begin to chase extensions offshore from platforms. The Soviet Union had ample onshore supplies, which meant that it had no particular incentive to invest in offshore drilling equipment. The Caspian itself was therefore largely left fallow, although the borderlands were thoroughly investigated. Of particular importance was the discovery of the Tengiz Field in 1979 in the prolific pre-Caspian basin of Kazakhstan, only some 70 kilometers from the shore. Silurian source-rocks had charged a carboniferous reef reservoir at a depth of about 4,500 meters beneath an effective seal of Permian salt. Initial estimates suggested a potential of about 6 Gb, but the problem was that the oil has a sulphur content of as much as 16 percent, calling for high-quality steel pipe and equipment, not then available to the Soviets. Development was accordingly postponed.

The fall of the Soviet regime in 1991 opened the region to Western investment. The oil industry in particular was enthusiastic that here might be a "new frontier" to offer them another lease on life, having effectively lost the Middle East through expropriation, and having thoroughly explored the rest of the accessible world. A glance at the map of the unexplored Caspian Sea surrounded by oilfields was enough to capture the attention of Western strategists, especially in Washington, who began to hope that in the Caspian they could find an escape from the stranglehold of the Middle East in their desperate quest for access to a foreign oil supply. These notions and ideas soon gained a momentum of their own, far removed from any thorough scientific analysis. There were many motives to exaggerate the prize, as strategists sought to shift foreign policy and mobilize military capability. Before long the Caspian had won the image of being a second Saudi Arabia, floating on oil.

A second look at the map reveals that it is not easy to get the oil out of this landlocked area, remote from Western markets. But this was manna from heaven for various geopolitical "experts," who could now dedicate their think-tank efforts to designing devious strategies for controlling the transit countries and building pipelines, of course taking the claimed geological potential for granted. As many as eleven schemes were considered, each with different obstacles. The obvious route was through Iran, but this would have given Tehran a critical control of future US supply, which was not thought desirable. Another was to the Black Sea, for shipment through the Bosporus in tankers, but the Turks objected that this would be an environmental catastrophe waiting to happen. Existing pipelines through Russia could be used and expanded, but that gave the Russians critical control. The Chinese too, who recognize their desperate dependence on growing imports, entered the scene with a proposal for a pipeline in their direction. Then there was Afghanistan, and a proposal by the American company, Unocal, for a gas pipeline from Turkmenistan to the Indian subcontinent. Another heroic idea was to pipe it to the Black Sea for transhipment to Bulgaria, for whom environmental issues are not a particular priority, and then into another pipeline constructed through the Balkans to the Adriatic coast of Albania, passing through Kosovo, again a route not without its hazards.

The preferred route at the time of writing appears to be overland through Azerbaijan, Georgia and Turkey, which would be no mean undertaking, having to cross high mountain ranges occupied by disaffected Kurds and others, who might find it a ready target.

In any event, the US began to establish and expand its military presence throughout the Caspian region with various bases and military "aid" projects in the nearby countries, and later toppled the Afghan government after a short campaign of bombing supported by ground forces of the Northern Alliance. The new President, Hamid Karzai, was himself associated with the Unocal pipeline project, directly reporting to Dick Cheney, and this project has now been resurrected. Meanwhile, enterprising Irish entrepreneurs import Caspian oil to Iran, re-exporting Iranian oil in exchange. All of this can be seen as a kind of replay of the so-called "Great Game" where, in the nineteenth century, various Western powers and Russia vied with each other for influence in Central Asia. However exciting this may be, it now begins to look as if the Caspian may not live up to expectation, as ten years of exploration and development by Western companies reveal its real and modest potentials.

BP took a pioneering role with Statoil, its Norwegian partner, when the Caspian opened. Interest was at first aimed at the offshore extensions of the Baku trend, where a number of prospects, already identified by the Soviets, were successfully tested, leading to the development of the Azeri, Chirag, and Guneshli fields. Some 17 "wildcats," as exploration boreholes are colorfully termed, have been drilled since 1992, finding some 3 giga-barrels (Gb) of oil, which while useful, is not enough to have any particular world significance. BP also investigated the Tertiary deep to the south, finding the Shah Deniz, a gas-condensate field. Evidently high temperatures on deep burial have broken down the oil into a gas, containing a high dissolved liquid content, as might be expected. This area verges on waters claimed by Iran, and seismic surveys have been halted by Iranian gunboats. It is significant that Russia's LUKoil, which was a partner in the Azeri–Chirag–Guneshli fields, has decided to sell out to a Japanese company desperate for access to oil; meanwhile ExxonMobil has withdrawn from Azerbaijan altogether. Evidence to date suggests that Azerbaijan has reserves of about 12 Gb, and since the larger fields are almost always found first it is unlikely that new exploration will bring the total to more than about 15 Gb, if that.

Kazakhstan also soon attracted serious interest. The American company Chevron (now Chevron Texaco), together with ExxonMobil, agreed to develop the Tengiz Field. It faced many operating and technical challenges, but has managed to build production to about

250,000 barrels/day, which is exported partly by rail and partly through the Russian pipeline system. One of the problems has been the disposal of the massive amounts of sulphur that have to be removed from the oil by processing. Plans to increase production by 480,000 barrels/day by 2005 have now been shelved, adding another nail to the "Caspian bonanza" coffin.

The greatest interest of all, however, attached to a giant prospect, termed Kashagan, which was identified in the shallow waters of the northern Caspian off Kazakhstan. Like Tengiz, it relied on a high-sulphur Silurian source, deep carboniferous carbonate reservoirs, and Permian salt seal. It had a huge upside thanks to its sheer size, offering a certain potential to become perhaps the world's largest oilfield. Jack Grynberg, the well-known New York promoter, managed to strike a deal with the Kazakh President, leading to the entry of a largely European consortium, comprising BP Statoil, the Italian company Agip, British Gas, the French company Total (now Total Fina Elf), and minor American interests. Grynberg retained for himself a so-called "overriding" royalty. But the initial enthusiasm waned when the companies began to get into the details. In geological terms, there were uncertainties whether the reservoir would be one large platform, or would turn out to be made up of individual reefs separated one from another by rocks lacking porosity and permeability, as experience from Tengiz would suggest. Seismic surveys showed that the integrity of the salt seal was weak in parts of the structure. The companies also found that they faced monumental operating challenges: the waters were shallow, making it difficult to bring in and position equipment, while also posing environmental threats to the breeding grounds for sturgeon shoals supporting the Russian caviar fisheries. If that was not enough, a gruesome, chilling wind blows in winter covering everything in ice. Nevertheless, the companies have succeeded, at astronomic cost, in drilling three wildcats, on what presumably are the most favorable parts of the prospect, announcing that they had found between nine and 13 Gb. BP Statoil decided to withdraw, exposing themselves to a lawsuit filed by Mr. Grynberg, who was not pleased to miss his overriding royalty. This is another big nail in the coffin, although the remaining companies, now led by Agip, soldier on.

In addition to these main projects, the Russians themselves have made a 2 Gb discovery in the northwest part of the Caspian, and Turkmenistan has announced an oil discovery of uncertain size off its mainly gas-prone territory.

In short, it has now become clear that the offshore Caspian has been a great disappointment. Exactly how much of a disappointment is hard to say, as the oil statistics are even more unreliable than usual. Total reserves for the offshore probably stand at about 25 Gb, with new exploration offering potential for perhaps another five, a good deal less than the 44 Gb mean estimate proposed in a study by the USGS in 2000 (and vastly less than oft-publicized wild estimates, extending up to "200 billion barrels" in the US and European press, through 2000–01).

Offshore production today is mainly confined to Azerbaijan, where it probably stands at about 250,000 barrels/day. Given the withdrawal of the major companies, the monumental technical and operating challenges, the uncertain contractual regime and export pipeline obstacles, it is difficult to be sanguine about the future rise of production. Realistically, it seems doubtful if it will be possible to reach a maximum of more than, say, about 1.5 million barrels/day in ten to fifteen years' time. If this plateau was achieved, and it would be effectively constrained by pipeline capacity, it might last another ten years before the onset of gradual decline at the then depletion rate.

The US currently imports some 11.6 million barrels/day, approximately 60 percent of its consumption. If demand were held static by recession or government policy, imports would still have risen to about 17 million barrels/day by 2015, in the face of the continued decline of indigenous supply. About 10 percent of its needs could come from the Caspian offshore, in the unlikely event that it was able to have exclusive call upon it. And even that would last only for a few years.

The foregoing discussion relates to the offshore Caspian, which seemed to be a particularly promising area, not having been explored by the Soviets. The surrounding onshore territories were thoroughly explored, so that most of the prospective basins and the larger fields within them have already been identified. There is naturally scope for more exploration and development, leading to production growth in the future, but that is another story. It is very evident that the Caspian has proved a chimera, dashing hopes that it would lessen US dependency on the Middle East. This realization perhaps explains in part why it now turns its guns on Iraq. There is at the same time a serious lesson to be learned: all that glitters is not gold. When the dust settles, Iraq may also be found to be able to offer less than was at first hoped, nature being immune to military intervention.

7
Update to the Caspian Chimera

Sheila Newman

Colin Campbell's comprehensive chapter, written in 2002 for the first edition, describes probable outcomes so well that there is almost no need to update it.

Kashagan: Recent descriptions of conditions in the field of this "super-super giant" sound as if it is located on a strange and hostile planet. The importance of the proportion of sulphur in different oils derives new meaning when you understand that to survive with the 15 percent concentration of incredibly toxic hydrogen sulphide gas in Kashagan oil, workers need to wear oxygen tanks and special suits, use special vehicles, and operate from specially built artificial islands. The oil deposit is located at 4.02 kilometers or 2.5 miles below the seabed at pressures around 500 times sea level.[1]

Not unexpectedly, the project continues to be marked by ever-rising costs and bureaucratic twists and turns. An increasingly assertive Kazakhstan government blocked a sale to China by the exiting British BG and bought half of BG's investment in the consortium, with the other half going back to consortium partners. But on August 20, 2007, China National Petroleum Corp. (CNPC), the biggest Chinese oil producer, announced its intention to expand oil and gas in cooperation with Kazakhstan in a 750-kilometer extension of the Atasu–Alasankou oil pipeline. The pipeline will, if completed, connect China with Kenkiyak and Kumkol oilfields, which are operated by CNPC in Kazakhstan. CNPC stated that it expects to obtain about 5 percent of its current requirements from the pipeline – 400,000 barrels a day.[2] On August 27, 2007, the government suspended ENI's[3] permit to operate in the field for three months, alleging numerous problems including breaches of safety regulations and poor environmental compliance causing threats to local species.

The Shah Deniz project finally produced gas in December 2006, had to close in January 2007, reopened for a short time, then closed down again with no definite reopening date, causing Georgia to buy emergency gas supplies from Russia at a very high price. Georgia would prefer to be much more politically and energetically independent of Russia and hopes Shah Deniz will assist this policy.[4]

Azeri, Chirag and deepwater Gunashli: Gunashli is not scheduled to begin production until 2008–09. Chirag has been producing since 1997. Central Azeri began producing oil in February 2005.[5] The problem of getting oil out of the Caspian has been solved in one area by the opening of the second longest pipeline in the world, the Baku–Tbilisi–Ceyhan pipeline, in July 13, 2006. Crude is now transported 1,760 kilometers from the Azeri–Chirag–Gunashli oilfield in the Caspian Sea to the Mediterranean Sea. The pipeline goes through the capital of Azerbaijan, Baku, Tbilisi in Georgia, to a Ceyhan port on the southeastern Mediterranean coast of Turkey.[6]

Iran's profile in Caspian oil exploration has been insignificant for about ten years.

NOTES

1. Guy Chazan, "In Caspian, Big Oil Fights Ice, Fumes, Kazakhs," *Dow Jones Newswires,* August 28, 2007, www.rigzone.com/news/article.asp?a_id=49533, accessed April 10, 2007.
2. Henry Myer, "China, Kazakhstan Link on Caspian Oil Pipeline," *Shanghai Daily,* August 20, 2007, www.shanghaidaily.com/sp/article/2007/200708/20070820/article_327883.htm.
3. ENI stands for the Ente Nazionale Idrocarburi (Italy), the National Hydrocarbon Organization.
4. Terry Macalister, "More Trouble for BP as Gas Scheme is Halted," *Guardian,* February 5, 2007, www.guardian.co.uk/oil/story/0,,2006007,00.html?gusrc=rss&feed=24.
5. Statoil home page, "Azeri, Chirag and Gunashli," www.statoil.com/STATOILCOM/SVG00990.nsf?opendatabase&lang=en&artid=751ECB67B89365C4C1256F16002A2A86, accessed August 30, 2007.
6. "Baku-Tbilisi-Ceyhan Oil Pipeline," http://tinyurl.com/2twwp5.

8
Battle of the Titans

Mark Jones

What is the nature of the current crisis? Where is it heading? What are the possible outcomes? The world geopolitical and economic system is holistic, and in many ways hyper-centralized and extremely fixed in how it behaves, but is composed of discrete elements with differing degrees of partial or relative autonomy. Different regions are subject to different dynamics. The rates of growth, or relative or absolute decline, differ between them. The system changes as a whole, but change also occurs in the existing equilibrium, and in balancing factors of regional economic and political power.

A.G. Frank, Immanuel Wallerstein and others have argued that the main trend in the world today is the decline of Anglo-Saxon hegemony and the re-ascent of Asia, and above all China. Legitimate questions arising include: Will this be a new American century, or is US hegemony as profoundly challenged as many now argue? Will China achieve regional hegemony, and is it capable of going on to true global hegemony? Or will China collapse when the sources of growth (easily identifiable and not the result of magic) fade, and underlying demographic, ecological, resource, and inter-ethnic strains start to tell, possibly destroying the unitary Chinese socio-economic space, just as the USSR was destroyed by its failure in global competition?

A second group of questions: If there is a transition going on from the global hegemony of American to Chinese dominance, and a transition from the present Anglo-Saxon world system to a differently ordered world with Asia as its center of gravity and propulsive dynamism, how will this transition occur and become effective? Is a world war thinkable, or can a peaceful transition take place from the Anglo-Saxon-centered world to a Sino-centric world? Can such a transition happen at all? Might the two systems abort in an endgame resource war for declining oil reserves, through uncontrollable climate change, or other factors?

It should be borne in mind that the changeover from declining to ascending hegemony can happen – and has historically – not by means of war but with the consent and active participation of the declining power, which in some cases may even push forward the claims of the ascendant power, surrendering its own hegemonic status and freely giving up many strategic and economic possessions. There are examples of this from antiquity and in the Middle Ages, and the modern example is that of the surrender by Britain of its global-hegemonic status to an initially unwilling, isolationist US during the 1930s and 1940s. This process did to some degree contravene, or at any rate qualify, Lenin's thesis about inter-imperialist rivalry always leading to war, although arguably it was Lenin's own success in creating the USSR which caused this, by forcing the British to take a defeatist view of their own prospects. Is it thinkable that the US might surrender hegemony to China? Maybe it is more thinkable than we realize, once we canvass the alternatives, and once we look beyond the rabid posturings and imperialistic breast-beating of the US ruling class.

It is salutary to compare modern attitudes with those that were prevalent in Victorian Britain, during the period of unquestioned British global supremacy. Check out Rudyard Kipling or J.G. Farrell – the British were at least as sure of their manifest destiny, of their imperial, civilizing mission, and at least as arrogantly confident about an empire on which "the sun would never set," and about the racial superiority of their kind, as is the US today. Nevertheless, the time was not long before the British abandoned imperial pretensions, packed their colonial kit and left. Winston Churchill, in his desperate attempts to lever the US out of its isolationist neutrality in 1940, gave away many key strategic assets to Roosevelt, including not only the British and South African gold reserves, but the global network of island bases on which British imperial communications systems depended (this was the backbone of the information systems supporting the world markets of the time).

The middle decades of the last century might best be seen as an interregnum, during which time the declining and ascendant imperialist powers, Britain and the US, colluded and collaborated to fight rivals (Japan, Germany) and marginalize or contain rivals (Bolshevism). Once US hegemony was assured (by 1945) it was relatively straightforward to restructure the global system on a new, US-dominated basis, and to create the institutions and frameworks for global commerce and international law that secured unchallenged US hegemony. Except for two brief periods during the Korean and

Vietnam Wars, the USSR ceased to be a serious challenge after 1945, despite the much-vaunted menaces of the Cold War period.

The British surrendered their empire in pursuit of their own best interests, and indeed for national survival. But the psychological trauma and the bitter taint of defeat scarred a whole generation of people, not merely in the British ruling elites but among wider social classes with a sentimental interest in the empire. This included wide sections of the British working class. We can especially identify the so-called aristocracy of labor, which shared most of the racist assumptions behind the ideology of empire, and benefited materially from the so-called "social imperialism" of the military Keynesian/welfare state reforms of the early postwar period, partly financed by, and riding on the back of, ebbing imperial wealth.

In its heyday the British Empire was more powerful and effective than the US Empire has been or is today. The British not only moved populations around in huge numbers; they also moved plant and animal species. The British did more to shift and transplant alien flora and fauna from one continent to another, thus reconstructing whole ecosystems, than any other empire, although the Romans did a lot more in that sphere than most people realize. It takes a lot of arrogance and certainty to do the kinds of things the British ever-so-freely took it upon themselves to do. They reshaped whole continents, from Australasia through Africa to Latin America. Successive waves of emigration from the homeland created a whole English-speaking world, of which the US was at first only a subset and which it finally inherited, but had not created. The British plowed their way through every precapitalist social formation they encountered, and either wiped it out or, through colonialism, totally reconstructed it. Yet despite (or because of) these grandiose achievements, the British Empire, which seemed so enduring, was a short-lived thing. There is nothing to suggest that the seeming permanence of the US imperium should be any less fleeting. There is also no reason to suppose that sheer self-interest might not drive Americans themselves into a recognition that the price of domination, alone and unchallenged, is too high, and that an accommodation must therefore eventually be made with a rival who might one day become a successor. Since the 1939–45 war, the US has in fact made a practice of co-opting present and potential rivals into junior partnership. It has done this not only to Britain, but also to Germany (1960s), Japan (1970s and 1980s) and latterly even to Russia (from 1991).

The US is now functionally locked in to Chinese industrial capitalism; the two states are in a tight embrace, performing a minuet which is part dance of death and part marriage of convenience.

Both states face many common problems, and each needs the other because of complementarities and useful asymmetries. However, the radical contradictions between them ensure that they are imperial rivals. And the balance of power between them appears to be changing. US economic and therefore military growth significantly declined from the later 1980s through to 2002. China has been growing faster, and has now entered a decisive phase of industrialization, where its industry is so diverse, deep, broadly based, and synergistic that it appears to be crossing a threshold, and is emerging as the world's premier industrial power, eclipsing all others, with an R&D capability equivalent to that of the US or Japan. Many estimates indicate that Chinese industrial production will outstrip that of the US during the present decade. It seems clear that Asia as a whole is now poised on the cusp of precipitous changes, and that the US is now clearly the declining regional power, giving way to rising regional hegemony for China. It is surely arguable, or even likely, that China could become the dominant power in Asia without the need for war, and with the US being unable to prevent this. Once the economic facts are in place, can the geopolitical consequences be far behind? What can prevent the binding together and fusion of Chinese, Japanese, and Taiwanese industry and capital, under Chinese hegemonic control?

Thus, for the first time in its history, the US now faces the distinct possibility of the partial eclipse of its global power. I cannot predict what will happen, and I doubt whether anybody can know, but it is surely plausible that the US will be obliged to accept its strategic defeat with good grace, to accept a seismic change in its status and position, because the US is simply no longer powerful enough to resist. The rise of China to at least regional Asian ascendancy seems to be already in the script, an inevitable and unalterable outcome of present trends.

Lots of caveats and objections, surely, can be entered here. For one thing, it may really be true that US technological and military supremacy is now so great that it will be impossible for China ever to "effectuate" its latent regional supremacy, and take advantage of its potential power during some future world crisis. One cause, but also consequence of this would be the world economy entering a very protracted period of stagnation and decline. Also, China may be plagued by crises of national origin on one of a number of fronts: demography, environment, resource depletion, and so on.

A crucial indicator will be which state or region comes out of a depression first, and A.G. Frank rightly dwells on this. Under present circumstances, it seems unlikely that global bourses will recover very quickly, and the present bear market might be very prolonged. In fact, any protracted bourse crisis on Wall Street, with the index hitting even 4,000 points, will trigger panic equivalent to the 1929 crash, the post-1929 collapse of the fantasy paper economy initiating a six-year, worldwide depression in the real economy. Optimists argue that, even in the event of a paper-economy meltdown, this will not harm the real economy because mistakes made in the 1930s will not be repeated. That is, protectionism, futile and counterproductive attempts to balance budgets, monetarist rivalries, the deflationary gold standard, and so on. Wrong. They will be, and in any case even if enlightened Keynesian policies are in place, it may not help. Keynesian deficit spending to bolster the domestic economy has not helped Japan in the past decade of increasingly futile attempts to spend the country out of deflationary recession.

The same people who say that policy and regulatory improvements mean that another 1929-style crash cannot happen also often said there would never be another bear market, and that the New Economy was a "paradigm change." None other than Federal Reserve chairman Alan Greenspan himself became an eager convert to the neoliberal view that soaring stock market index numbers were not, as he first thought, the result of "irrational exuberance" but represented a fundamental change in the economy and, in particular, a quantum shift in the rate of productivity growth. But we shall show that a slump can happen whatever kind of Keynesian demand management is attempted. Deficit spending does not overcome modern deflationary crises. So, as Wynne Godley argued in the London *Financial Times* in late 2002, a real and perhaps catastrophic slump, a real meltdown of the US economy in particular, is now a distinct possibility, even a probability. In that case, we might get a near-decade of mass unemployment in the capitalist heartlands, and a deepening pauperization of the peripheries. Yet this will take place on a geopolitical scene in which *the essence of the epoch is long-term competition for supremacy between China and the US.*

As Henry Liu argues, China abandoned its decades-long attempt at autarkic development during the Mao Zedong years. Instead, it has elected to join the world market and build a modern industrial system on the basis of export-led growth, rather than through self-sufficiency and import substitution. This is a highly dangerous strategy for both China and the world, although probably inevitable – China's leadership had no choice but to break out of the Maoist policy dead-end. It is dangerous for China because it entails an unsustainable commitment to growth through exports. Classical trade policy explains why overdependence on exports and capital inflows from abroad leads to further impoverishment of the masses, social tensions, and the inability to renew and develop essential social and economic infrastructures. At the same time, aggressive exporting helps destabilize the world economy. The counterpart to huge Chinese balance-of-payments surpluses are the increasingly unsustainable US trade and finance deficits. Additionally, aggressive exporting is inevitably deflationary. In effect, it pits the Chinese working class against the working classes of rival states. The question then becomes: Which of the rival, classic capitalist states is best able to raise its domestic exploitation rate? The winner will bankrupt its competitors. If successful, the strategy stands a good chance of destroying the bases of US world hegemony by destroying the US economy through endless deflationary down-spirals, starting with a savage cutback in overblown equity values. Chinese economic development policy is therefore – effectively, if not consciously – a policy of imperial rivalry and confrontation targeting the US.

The region which emerges first from a major and prolonged slump resulting from deflationary competition (and a slump is now surely on balance more likely than not) will be well placed to move out of mere regional dominance, and take its chances at becoming the unrivaled world superpower. If China survives the shocks and strains caused by a global slump in demand, then it is certainly well placed to emerge first from the subsequent trough. It is this line of thinking which leads me to argue that we probably are, as Frank says, in the throes of transition between hegemonies, and that indeed the present world economic crisis may itself be symptomatic of this crisis of transition, just as global collapse in the 1920s was symptomatic of the final decay of British power. China is more competitive, and this is the bottom line. Conventional or classic recovery for the capitalist world economy is likely to occur, at least through 2010–15, from any economic setback. The power with the underlying competitive edge will come out first. It will then be positioned to begin the process of institutional, legal and commercial restructuring to entrench its hegemony and ensure its dominance.

There is, however, one huge caveat to this whole line of argument. It does not take account of underlying, apolitical, extra-human, and planetary limits, which will most certainly afflict the world system, and which mean we are entering a still more radical and decisive epoch of historical challenge, upheaval and transformation in everything from geopolitics to everyday life. As is indicated by the very title of this book, no geopolitical analysis can ignore the colossal threats posed by the storm of self-reinforcing crises – anthropogenic climate change, mass extinction of species, and destruction of the biosphere – all of which were enabled by fossil energy supplies, whose wipeout will be rapid. These factors form a vast backdrop to all world-system or economics-based considerations of inter-imperial rivalry à la Lenin. But before attempting to integrate this domain of issues into the discussion, let us consider again this central question: What if China emerges first from the probable imminent global economic slump?

This slump, if it occurs, will hit the US and Europe especially hard. The dollar will decline, industrial output will shrink, consumption levels and living standards will drop with incomes. The US economy will be first and hardest hit, notably because it is wildly unbalanced, extremely dependent on cheap energy, and equally dependent on the dollar's "reserve currency" hegemony, which enables its payments deficit to be ignored. Any US economic slump will inevitably bring with it a collapse in personal and public consumption levels. Obviously this collapse in US demand will hit major exporters, China above all. China will then have to find other outlets for its huge industrial output. This means exporting to other regions which may be doing little better than the US, such as Europe and Japan, India, Southeast Asia, Russia, and the Latin American countries – but perhaps firstly, countries which export oil, minerals, metals, and agrocommodities, which profit from higher prices for these items.

The major weakening of the dollar (if it happens) may take the US out of the current game for a very long time. We will then have a situation in which the US, too, must export its way out of trouble. Now we shall really see just how competitive the New Economy is, and how much US productivity really increased in the "dot-com" years. Under any hypothesis, however, no sane person would bet that the US can beat China at its own game. Walk around your house and mentally eliminate everything made in China, and see what's left. Now try finding anything with "Made in the USA" on it.

Once you strip out dollar hegemony and the advantages of being the global reserve currency (or "currency of last resort"), you are left with a very naked emperor. Except for weaponry, control of the skies and sea lanes, control and surveillance of world information networks, and a powerful propaganda machine, the US has few cards left. Take away dollar hegemony and you have just another regional economy with its fair share of internal problems (soil exhaustion, aquifer depletion, near-exhaustion of domestic oil and gas reserves, lack of alternative energy supplies, a polluted environment, poor infrastructures, badly designed and expensive-to-maintain urban environments). If the US has to compete on a level playing field with the rest of the world, then it may find that its urban infrastructure is just as uneconomic and unsustainable as was the Soviet Union's loss-making effort to base itself on the industrialization of the Urals and Siberia. The US currently uses twice as much energy and raw materials per capita as the EU15 average, and more than ten times that of China. It is desperately uncompetitive. When the dollar has to be backed up by real values, US per capita GNP may fall by half in just a few years, as in the Great Depression. Under these conditions it is hard to see how the US can hope to maintain its global reach and present hegemonic position.

Since China faces similar global resource, energy and environmental challenges, and since the collapse of the world market must increase internal social instability, the Chinese regime will not wait around for the US to put the world to rights. It must produce – and export – or die, as outgoing President Zhou explained was the only choice for China during the 1997 Asian monetary crisis.

The two states who first came out of the Great Depression were Germany and the USSR – and each broke free through massive military spending. More recently, the Reagan "economic miracle" of the 1980s was largely driven by vast military spending, financed by government borrowing. Undoubtedly this is the first option large, militarized states consider when their leaders grope for reflation, lowered unemployment, and re-emergence, with yet more power, on the world scene. A program of "military Keynesianism" is certainly an option for China, which is just beginning to expand and modernize its military massively. American defense spending is by comparison much less sustainable at present, let alone projected levels, because of skyrocketing trade and finance deficits, which will be intensified by a weakening dollar. Moreover, the extremely capital-intensive nature of US weapons programs means that defense spending does little to galvanize the wider economy. This is not yet true of China. Therefore, in a major world depression China could probably increase its national military spending dramatically, knowing that this will boost its economy while reinforcing its drive towards hegemony. The Chinese navy already carries out much gunboat diplomacy in the coastal states of Asia. The Chinese will surely seek to emerge from a depression through military and economic domination of the entire Asian region, by saturating markets and hegemonizing its skies, seas, and data networks.

Sino-American joint or shared global hegemony may be the strategic compromise both states will entertain, for want of an alternative. Neither really wants war, but the resource and energy imperative may force the hand of either party – more likely that of the US. Under any hypothesis, however, the US will have declining hegemonic power, and China's will increase.

The last ten years have seen the greatest unforced capitulation in history – the uncontrolled implosion, or unconditional surrender, of the USSR, exploited by the biggest pyramid scheme in history (as Wynne Godley argues), all helping to create the biggest stock market bubble in history. The present bear market is not just a correction to that unprecedented human folly, unless you call Alaric the Hun's visit to ancient Rome a simple tourist's jaunt. This is surely the beginning of the end, not just of equity-culture, but of global Anglo-Saxon suzerainty. A.G. Frank was right: the pendulum is swinging back to Asia, but it is doing so under the final blowout of the model of petro-capitalism.

If, on the other hand, we are set on a course of global war, which was the outcome for "classic" economic depressions before 1914, and again through 1929–36, then Americans have only a very small window of opportunity (like Hitler enjoyed in 1939) before their military advantage evaporates. This is perhaps the real cause of Bush's headlong rush to war. It is China they must pre-empt. The Islamic world, broken-backed as it was and remains, is not the problem. This will be a war for the survival not only of the American Century but of that cultural zone where people actually live – the "burbs" with an SUV in every drive – the pinnacle of ostentatious consumption, born and raised on cheap oil.

Even in his lifetime, Lenin recognized that his original assumptions about the necessity of inter-imperial warfare and the certainty of subsequent proletarian revolutions would have to be qualified in light of new realities. When the science-fiction writer H.G. Wells visited Lenin in his Kremlin office, Wells told Lenin that although he did not know what weapons the next war would be fought with, he was quite sure that the one after it would be fought with bows and arrows. Lenin did not disagree: it was already apparent, before the advent of nuclear weapons, that modern warfare imposed intolerable costs on civilization. It was this realization above all which prompted Lenin's notions of peaceful coexistence: What use would the proletarian revolution be, if it inherited a wasteland? The effects of civil war and war of intervention on the young Soviet Russia from 1919 to 1921 was almost as catastrophic as nuclear war. And the USSR never recovered from World War II, as Mark Harrison has shown in his admirable studies of Soviet economic development. The propensity of the bourgeois to overconsume and destroy the earth, rather than let the workers inherit it, presents a conundrum which neither Lenin nor any other revolutionary has successfully addressed. But in the twentieth century the bourgeoisie also learned a terrible lesson. While we should not assume that war between the US and China is inevitable, we can be sure that the dynamic of history shows that the decline of US hegemony *is* inevitable, and that China will be the beneficiary.

It is clear to both these powers, and any interested observer, that the keystone of US global power is Middle Eastern oil. This was true throughout the last century and is even truer today, as the US confronts a proven domestic oil reserve base of around 30 giga-barrels (Gb), and consumes 6.5 Gb per year. The US energy crisis is both structural and accelerating. In the short to medium term these energy supply difficulties might be met partly by conservation measures, because the phenomenal wastefulness of US society leaves much scope for saving. But this is not necessarily compatible with robust economic growth despite the baying of Amory Lovins on the virtues of the unproven hydrogen economy. Above all, the US cannot afford to lose the economic race with China that at present it *is* losing. As I have already said, China's gross industrial output will probably exceed that of either the US, Japan, or Europe this decade. This will leave military control of Arabian oil as the remaining strategic asset – together with some military, intelligence and electronic technologies – to shore up the US global position. The US's pre-emptive move on Iraq will be largely designed to pre-empt China from asserting its power in the Middle East and becoming economically and strategically dominant there too. Can this US strategy succeed?

Looking ahead to the next 20 years, can Chinese hegemony consolidate itself? This would bring us back what may be termed, à la Chinoise, the Five Great Evils: anthropogenic climate change, mass extinction of species, destruction of the biosphere, resource depletion, and exhaustion of cheap energy supplies. But the Five Great Evils have no meaning by themselves. Rather, it is how people change in response to them that matters, and this is determined in the form and intensity of class struggle. At the moment conventional class struggle has almost no remaining political form, as mass societies around the world converge in a race to outwit, outpace, or simply ignore, for a little longer, the gathering storm.

9
Dark Continent, Black Gold

Andrew McKillop

Since the return of "oil angst," from 2001 as its price steadily mounted, the chancelleries and powerbrokers of the old Great Powers – the US and European countries – plus deciders in the emerging supergiant powerhouses of the globalizing economy, China and India, have specially pinpointed Africa. Africa has since then retained a special position in the crosshairs of political, economic and industrial initiatives by the world's power elite, with one objective: get to Africa's reserves of black gold, that is oil, before the others. From around 2001 and 2002 a flurry of initiatives, state visits, studies, projects, and actions have underlined the growing importance of Africa for European, US, and Asian energy importers – if only to assure some distribution of risk away from the present and potential battle zones of the Middle East, and if possible to get Africa's oil a little cheaper than elsewhere. One example of this is the World Bank "anti-corruption and good governance" initiative of Paul Wolfowitz in Chad, delivering Chadian oil, produced by ExxonMobil, to the US at around US$20 per barrel below the world price.

Elsewhere in the oil-pumping community, countries such as Iran and Venezuela, as well as Nigeria in the Dark Continent, have experienced brewing civil conflict, in the Iranian and Venezuelan cases leading to the menace in theory, but not practice, of US invasion "to restore democracy." Elsewhere in Africa, and especially in Sudan, oil geopolitics and the big consumer nations' "quest for human rights and democracy" are easily melded into permanent invasion threats. On the world stage, emerging Great Power rivalry for oil reach is a constant reminder of how coming peak oil presages future and *permanent* shortage, followed by almost certain standoffs and conflict between the old Great Powers, led by the US, and the new Great Powers, led by China and India. The common thread and theme is their need to get oil.

A few simple facts and figures underscore the oil interest of the Dark Continent. Due to extreme poverty and underdevelopment, African nations consume a tiny fraction of the per capita oil used by the urban industrial, developed nations (see Table 12.1, p. 144). Africa's oil production in mid-2007 stood at around 10.5 million barrels/day (Mbd) while consumption was only 4 Mbd, leaving 6.5 Mbd to be shipped out to world consumers of plastics and petrochemicals, and drivers of the latest "energy efficient" cars in the advanced nations. Other than Saudi Arabia and Russia, no other oil export supplier can match this volume. Frequent press reports in finance-oriented journals like the *Wall Street Journal* or *Financial Times* never fail to point out that Africa now "outproduces the North Sea," for one reason because North Sea production is falling[1] at about 6–11 percent per year and Africa's relatively small oil reserves have only just started being exploited. The actual reserve size and geological prospects of the West African region are in fact modest. Indeed, West Africa, both onshore in Nigeria and offshore, particularly Angola, has so far proven reserves of mostly offshore, deepwater oil that amount to less than 4 percent of world proven reserves, or around 35 billion barrels. By comparison, the North Sea at the start of its rabid, wildfire production history, in 1980, contained about 55 billion barrels. West African oil and gas discoveries in recent years are in line with this slim, restricted, difficult-access resource base; any *additional* production from the region in the 2010–20 period is therefore unlikely to exceed about 2–2.5 Mbd or around 50 percent of current and declining output from the North Sea.

The central point is that only because oil consumption by *all nations* of the Dark Continent is so low, through poverty and underdevelopment, can Africa provide any hope of small "stopgap" supplies to the world oil market, pushing back the fatal reality of peak oil by a few months or years. In addition, increased local consumption is a reality, and the real short-term priority for African countries. Any serious attempt at conventional economic development in Africa will almost certainly lead to *reduced* oil export capacities, either rapidly, or within at most a decade.

Such facts of course do not worry sensation-seeking editors, and the eyes of oil-hungry OECD national leaders remain riveted on the Dark Continent simply because it is so dark. As the Light Pollution Institute and International Dark Sky Association websites (darksky.org and lightpollution.it) graphically show, where the night skies of Europe and North America are brightly electrified, the African continent appears like an inky black hole. A few figures show why this is so: for countries in the Ecowas (West African) group, including Black Africa's second biggest economy – oil-exporting Nigeria – average electricity consumption per capita is around 50–75 kilowatt hours per year, compared to annual average European, Japanese, or US consumption rates of *around 3,500–5,000 kWh/capita.*[2]

This situation is unlikely to maintain itself. Many upbeat business press articles cheering on new production *opportunities* from non-OPEC countries will add that much or most of any projected but unlikely "explosion" in African oil and gas production, almost exclusively on the western side of the continent, will be heavily oriented to offshore production. While ignoring the very high costs of this continent-edge exploration, discovery and then production in water depths already exceeding 2,000 meters or 6,500 feet (Angola holding the world record for extreme depth offshore oil production), these upbeat articles will sometimes note, discreetly, that being offshore these vital installations will or can shelter this *lifeline supply* from the many civil wars that have ravaged Black Africa since the 1980s.

To a certain extent it is possible to speak of a "pan-African war" ravaging the continent in the 1990s – to the splendid disinterest of today's champions of "humanitarian-based invasion" of underperforming and potential Black Africa oil exporters. The "hearths" of this war include the former Zaire, or Democratic Republic of Congo, but secondary zones of almost constant armed conflict include many countries in West Africa. The basic motor of this armed conflict is poverty, needing a dramatic change of policy and attitude in the rich world, and a vast increase in aid or the price paid for African raw materials and energy exports, and fast growth of local or domestic oil consumption – squeezing exports. Without this, the pan-African war can at any time reignite and continue. As a base for finding and producing cheap oil and gas this very surely creates "ambivalence" and real challenges – currently and predictably responded to by leaders of the so-called international community calling for more military intervention in African hotspots.

It would be no exaggeration to state that Africa's experience of "structural economic adjustment" imposed by the International Monetary Fund (IMF) and the World Bank, and condoned by the world's self-elected "defenders of human rights and democracy," led to Africa being more wracked, exploited, colonized, and oppressed in the period from around 1985 to 2000, than it ever was in the heroic times of slave trading and colonial war, either before or after the "Carve-up of Africa" through the 1850–1900 period, or in any ensuing liberation war. This onslaught by free market forces, tied loans, and bargain-basement priced commodities produced in Africa is well documented by writers such as Susan George and Naomi Klein, and was a grim *retour du baton* following Africa's brief day in the sun, during the hike in primary product prices triggered by the first oil shock, and lasting through about 1975–82, during which international lenders fell over themselves to finance huge projects for minerals, metals, and agrocommodity development throughout Africa.

With the Thatcher-Reagan recession and slump of 1980–83 these loans became close to impossible to repay. The IMF and creditor governments of Clinton's rich world imposed squeaky clean *balanced budgets*, policed to the last dollar, to the last kilo of food *not* imported and not fed to tens of thousands of undernourished children. By 1986 Africa counted a string of countries where more than 25 percent of total export receipts for their primary product exports, whose prices had fallen far and fast from their 1975–82 highs, were needed simply to pay *interest due* on "sovereign" or state-guaranteed loans.

Apart from some well-publicized debt forgiveness, Africa's unpayable debts are consigned to the rich man's Club of Paris[3] (including Russia in its management[4]), which takes over debt management after private banks throw in the towel over "non-performing loans" by debtor countries.

Today, funds controlled by cash-rich China and India, Russia, Kazakhstan, the UAE, Saudi Arabia, Singapore, and other *sovereign national wealth funds* are crowding into the African debt buyout arena – with oil as the key attraction, followed closely by metals and minerals resources.

The shameful impoverishment of the poorest countries and communities in the world not only transformed Black Africa's few oil and gas exporting countries into the most servile *price takers* that any yuppie economics guru could boast, but also prepared conditions for large-area, multi-country rebellions, massacres, and wars. Today's Darfur conflict and chaos in Somalia are just two examples, but when conflict threatened functioning and important oil exporters, specially Angola and Nigeria, all kinds of shuttle diplomacy and financial handouts have been used to protect the West's oil-supplying *performing assets* while continuous untold barbarity can go ahead in the *free-fire zones*.

This pattern of events in the Dark Continent is now under threat from an unexpected source: peak oil. Keeping the "environment friendly" cars of city commuters going, and the plastic bags of consumers full of petrochemical goodies in the now massively extended global economy makes it necessary to tone down and limit the carnage in Africa, some of which contains *untapped oil reserves*. Booming non-oil commodity prices now enable Africa to import not only massive supplies of small arms and ammunition, but also capital goods, machinery, and around 3 million secondhand cars per year, in 2006, from Europe and Asia, without crawling to the IMF for bailout loans under draconian conditions.

In the period from 2000 to now, policy and attitudes of the so-called international community towards Black Africa have rapidly changed, mainly because of dwindling world supplies of oil and gas. As shown by many typical, and farcically exaggerated, reports on black gold in Africa, oil from the Dark Continent is a hope or chimera reinforced by regional war and strife in the Middle East and Central Asia. Exactly as with the buildup to massive investment in the Caspian oil province, it is necessary to exaggerate reserve and production potentials. Unlike the Mid Eastern and Central Asian crises of today, this mix of oil greed and fear of losing supply sources is associated with the slowdown, and in some cases the stop, of the many *brush-fire wars* that previously flickered in Africa's pitch-dark night-time skies.

From 2001, Angola's offshore oil prospects and development programs rapidly expanded.[5] The bloody Jonas Savimbi, the "freedom fighter" so admired by Ronald Reagan, was quietly shot one day, signaling an effective end to a 26-year civil war. Angola's approximate 1.9 Mbd of exports, all drawn from extreme-depth offshore wells, are an increasingly precious *lifeline supply*. Similar concerns about Chad's 0.3 Mbd probably motivated the end of the three-horse race of Libyan, French, and US-financed killers in Chad's long-running civil war, in 2003–05. Black Africa, through extreme poverty and under-development, consumes small quantities of oil, making it possible to export impressive-looking volumes for the global community, and push back the date of peak oil reckoning.

Elsewhere, Libya's Algerian junta "rehabilitators" (after Libya's theatrical "abandonment of nuclear ambitions") control about 33 percent of European Union gas import and 20 percent of EU oil import supplies – and quietly continue their war with Islamic fundamentalists. This effective civil war, "benignly" neglected by media and politicians alike in the "great democracies," has claimed at least 350,000 lives. The Algerian generals who annulled the results of a 1992 election will rely on weaponry, training, and diplomatic support from their Anglo and European backers, while Algeria's declining gas and oil reserves hold out.

Diverse chaos was willed on the Dark Continent via an absurd, degenerate, economic doctrine and ideology, called New Economics in the better-fed nations, but *Belsen Economics* in Africa.

Since 2002, US permanent troops have been stationed in Equatorial Guinea, coinciding in 2005 with the doubling of this tiny country's tiny oil production – of which abjectly poor Guineans consume very little. In 2002, France established its SIGAfrique network with a €1.5 million grant to national geological surveys of several West African countries, to store, securitize *and re-analyze* minerals and petroleum resource data – to "safeguard the patrimony" of these countries, of course. For many years the Swedish government has funded development effort, now supported by the European Commission, to *increase fuelwood burning* in African countries[6] that might – dangerously – start to consume small amounts of oil, causing higher oil prices for Europeans. Since 2005, these and similar initiatives to *get at African oil* and to prevent Africans from consuming too much of it themselves have multiplied.

The rich world is wise – or blessed by geology – in focusing on Africa's offshore oil and gas, far from land and the danger of *damage to installations*. Africa has more basic challenges to resolve. Africa's entire existence and survival is threatened by AIDS,[7] the *certainty* of increasing war and civil strife, plus the crushing poverty which is a hereditary birthmark of Belsen Economics. This is the morality and humanity we can expect from the "inspired" creators of New World Order, yet even sacrificing all of Africa will not stall the arrival of peak oil.

NOTES

1. See Roger D. Blanchard, "The Impact of Declining Major North Sea Oil Fields Upon Norwegian and United Kingdom Oil Production," www.dieoff.org/page180.htm.
2. UN PIN (population); *BP Statistical Review of World Energy*; ES EIA; OECD IEA.
3. The Club of Paris, made up of 19 lender countries (the US, Japan, 13 EU countries, Norway and Switzerland, Russia, and Canada), has the remit to negotiate, strictly in private, conditions relating to public bilateral debt of developing countries and their refinancing. Club members hold about 30 percent of all outstanding debt on third world loans. The Club intervenes when a debtor country submits to the IMF's ultraliberal strictures. Source: www.attac.org.
4. The importance of Russian arms and war loans to Angola through the 26-year civil war (where the US backed Jona Savimbi's UNITA forces) is

that these are secured by Angola's *future oil production*, as are the loans provided notably by Swiss and French banks to the winning MPLA regime. Consequently, there is not a cent left over for development, and Angola's newfound freedom will be "celebrated" in perfect poverty. See also T. Hodges, *Angola from Afro-Stalinism to Petro-Diamond Capitalism*, James Currey, Oxford, and Indiana University Press, Bloomington, in association with the Fridtjof Nansen Institute and the International African Institute, 2001.

5. "ExxonMobil Starts Marimba North Project Offshore Angola," *Scandinavian Oil and Gas Magazine*, October 26, 2007, www.scandoil.com/moxie-bm2/by_province/africa/exxonmobil-starts-marimba-north-project-offshore-a.shtml.
6. SADC Energy Programs and Projects, http://tinyurl.com/2rs6hd.
7. Many African countries have over 15 percent of their population HIV-positive. By 2020 this rate could increase to an average of 25–35 percent. (Data from projects funded by the Global Fund to Fight AIDS, Tuberculosis and Malaria, Geneva, and Mr. Honoré Nkusu Zinkatu Konda, *Oeuvre médical au Congo*, OMECO, Kinshasa.)

10
The Chinese Car Bomb

Andrew McKillop

EXPLODING NUMBERS

Exploding numbers of oil- and gas-fueled road vehicles, including cars, buses, trucks, motorcycles, scooters, mopeds, all terrain "fun" vehicles, and agricultural and construction offroad vehicles,[1] draw much less attention than *human population numbers* as a "possible threat to continued wellbeing." This also applies to public and private aviation, where typical growth rates are even more unsustainable. Plane movements and numbers of new planes coming into service in the period 2005–10 are at, or forecast at, around 8 percent or 9 percent per year for civil aviation, and up to 25–33 percent per year for business jets. Forecasts for this last and booming sector with the absolute maximum carbon footprint are *very bullish*!

Studied ignorance of exploding land transport fleets and their very direct impact on world oil demand is curious, given the phenomenal growth rates we find almost anywhere in the so-called emerging economies. On a world scale this can be shown by a few simple figures: in 1939 the world's roughly 2.3 billion inhabitants shared a total of around 47 million motor vehicles. Today's 6.5 billion human beings have around 875 million land transport vehicles to fuel, repair, park and run, almost exclusively (about 98 percent) using petroleum and natural gas at this time – though of course running on cheap hydrogen and abundant corn ethanol, or other miracle fuels, "by 2035 or 2045."

Human numbers increased less than three-fold or 200 percent in the last 68 years, but the world's car population grew by well over 1,750 percent.

Today's human population is growing at ever lower rates, for a variety of rational and less easy-to-explain reasons, and the composite or world average demographic growth rate is now hovering at not much above 1.1 percent per year – compared to 3.5 percent per year at its peak in the 1960s and 1970s. This was enough, as we know, to almost exactly double the world's population from 1965 (3.25

billion) to 2007 (6.5 billion), but the next doubling either will not happen at all, or take 65 years or more. Today's *annual increment* or potential recruits to the oil consumer global society is now running at perhaps 70 million per year.

World new car output at around 63–65 million per year in 2006 and 2007 is therefore closing up very fast to the cornucopian ideal of *"For Every Baby a New Car is Born."* When we add in world production of motorcycles and scooters, about 25 million per year and already about 70 percent produced in China[2] and India, the ultimate cornucopian dream of *at least one road vehicle per human being* looks distinctly attainable – given unlimited oil supplies of course. Each car, to be sure, weighs a lot more than each baby, most of them born in poor or very poor countries. Each car also needs plenty of oil, gas, electricity, and energy-intensive materials to produce – unlike human babies. Above all, each car needs to fill up the second it hits the road. If the road doesn't exist – like in China or specially in India – then it has to be built using plenty of oil.

UPPER LIMITS

Ross McCluney[3] explains the ultimate Heat Limit on world human population numbers, a true and final limit on human numbers which we will certainly never approach, but there are set and final near-term limits on the possible growth of world transport fleets. The most real, and most denied, of these near-term limits is very simply *the world's ultimate reserve of petroleum,* unless we wish to fantasize along with former US Energy Secretary Spencer Abraham by giving any credence at all to his November 2002 statement that the world "will have a total of 3.5 billion motor vehicles by 2050."[4] If this fantasy fleet were to come about – adding around 2.6 billion more vehicles to the world's current stock – the fuel requirements for 3.5 billion motor vehicles, at current average consumption rates, *would increase world oil consumption by about 70 percent.* Fueling up this fantasy fleet would take about 55 million barrels per day (Mbd) on top of current total oil demand. Because it is simply impossible to fuel this fantasy fleet on oil and gas, Abraham went on to add – and all similar reality denial experts add – the world's Fantasy Fleet of the Future will "of course run to a large extent on hydrogen." The "large extent" was carefully not defined by Mr. Abraham, and is even more carefully not defined by the *tutti quanti* who spin likeminded fantasies. We can note that in 2002 Abraham was unable to invoke the miraculous new fuel – corn ethanol – which the US of G.W. Bush imagined, for a short while, was the closest possible thing to instant salvation!

THE ASIAN PATHFINDER – JAPAN

To set the ultimate limit for growth of *petroleum- and gas-fueled vehicles* we can start with the near-ultimate example of a "catch-up" country in the car business – Japan. Even as late as 1949, Japan's transport fleet still counted some 146,500 horse- and ox-drawn carts, compared with fewer than 200,000 trucks and about 100,000 private automobiles. But through a self-reinforcing, very *high gain feedback* process of growth, with annual growth rates typically of 15–20 percent, year in and year out, Japan's private car fleet explosively grew from these small beginnings to its first million in 1963, to 5.2 million in 1968 and about 26 million in 1982, and over 45 million today. Also today, Chinese and Indian car fleets are growing at very similar rates, around 15–25 percent per year, *but in vastly bigger countries*. In Japan, annual growth rates, due to simple saturation effects, considerably slowed by the 1970s and 1980s. *Fleet size effects,* however, still permitted this slowed growth to give impressive annual increases in numbers of new cars hitting the road.

Growth is good, all good consumers believe, and to get on the growth track Japan's administrative elite, even after the culture shock of atomic weapons use against their civilian populations, and military rule by US Governor MacArthur, had to throw off mindsets dating from the 1918–39 interwar period, when road vehicles were seen as simple "feeders" for short-haul transport to rail, canal, river, and coastal shipping points or *transport nodes*. Japan's domestic policy makers, at the start of the postwar period under US occupation, thus preferred to spend money on repairing and improving the rail, shipping, and public transport sectors. In addition, their policy view on downgrading road vehicles was reinforced by Japan's terrain, its dense urban centers, and by Japanese feelings of doubt on the safety of cars:

> Because of slow improvement of the country's narrow, often mountainous roads, the government tended to discriminate against motor transport on grounds of road safety. City streets were often dangerous too. There was strict traffic control, rigorous tests for driving licences and careful inspection of all new vehicles, both home manufactured and imported. A high standard

of maintenance was promoted and the manufacture of reliable, safe cars was encouraged.[5]

The date that Japanese transport policy switched to an outright pro-car policy can be set at around 1955–60. During this period, animal transport completely disappeared from the agriculture sector as Japan experienced "catch-up" economic growth from about 1955. Despite this, however, as late as the early 1960s, Japan's Economic Planning Agency continued to purposely underestimate forward growth of road needs for the exploding vehicle population.[6] Today, the ongoing "restructuring" of Japan's national railway corporation, like its US role model, amounts to an effective bailout by an underfinanced, neglected public rail transport system. World-famous for manually compacting passengers into wagons like sardines into tins, Japan's rail transport system is a dwarf compared to the road vehicle sector. As elsewhere in the "liberalized economies" of the aging Organization for Economic Cooperation and Development (OECD) group of countries, Japan's national passenger transport depends almost entirely on the existence of oil-fueled private road vehicles running on compressed natural gas and above all on oil – from motor fuel to oil-based asphalt and bitumen.

THE SECOND COMING

While the US and, perhaps surprisingly, New Zealand were the fastest growing countries for *motorization* in the entire period 1905–40, achieving ownership rates for private cars of nearly 300 vehicles/1,000 inhabitants in 35 years despite starting from a near-zero base,[7] their growth rates peaked out well before World War II. These countries, with all the older urban-industrial countries, however experienced a "second coming" from the early to mid-1950s. Countries such as Canada, Australia, Italy, UK, France and Germany, experienced a new car-ownership growth bulge in the 1950–70 period. Typical growth rates were around 500 percent in 20 years. The UK, for example, experienced a six-fold growth in car and private vehicle numbers through 1950–70.[8] At the time of the first oil shock of 1973–74, only Japan experienced a strong *but short* downturn in this "motorization" trend. From no later than 1975–80 the tried-and-tested "economic growth model" of car-based and car-oriented growth, a key concept in economic mythology from the time of Henry Ford in the US of the 1920s, was transferred and applied with full force in several nontraditional car-owning democracies and dictatorships of the time, including South Korea, Brazil, Malaysia, Turkey, Iran, and the Soviet Asian republics.

Somewhat later (towards the end of the 1980s), but with truly *unlimited upside potential,* China and India adopted this economic growth strategy. In a rapidly intensifying movement, since 2005, global car-makers have been flocking to sign joint ventures and international export sales deals with China's and India's car-makers – whose total output will likely exceed 10 million cars in 2007.

Today, in countries such as Germany, the US, France, and Australia, it is not difficult to find three- and four-car households, nor 25-mile tailbacks every weekday on every main highway into congested, sprawling, and polluted downtown centers. The same occurs in São Paulo, Bangkok, Ankara, Seoul, and Kuala Lumpur. Shanghai and Beijing are already close behind in the race to have daily crawl-ins on their fast-expanding autoroutes and urban highways. Such is progress.

The "oil demand multiplier" inherent in motorization is explained by a vast range of products arising from the magic of petroleum-based chemicals industries, notably plastics, composites, paints, and resins. All are essential to the modern private motor vehicle industry. As in Henry Ford's time – when animal bone and ligaments, skins, wood, and wood resins were still extensively used in car manufacture – the economic *multiplier impacts* of "unfettered growth" of the car industry remain highly attractive to economic planners, resulting in motorization continuing to spread out and away from the core countries of the aging "advanced industrial" OECD North. As the world's biggest car-makers know, saturation effect in the OECD North can be more than compensated for by the "unlimited growth potential" of the *Chinese car bomb.*

THE ULTIMATE LIMIT

There are, however, distinct limits on the ultimate reach of this tried-and-tested "growth strategy." Today's private car and similar vehicle or "light truck" ownership rate in the US is around 745 vehicles/1,000 inhabitants. Lower but similar rates (around 400–650 vehicles/1,000 population) obtain in Japan, South Korea, Italy, Belgium, Germany, France, the UK, and other car-saturated economies. Applying the same ownership rates to India or China, assuming these motor vehicles would, could, may, or might be oil- or gas-propelled, results in absurd numbers for annual oil or oil equivalent gas consumption. In the case of China's car fleet, we are already, using World Bank and other data, at the fantastic but real *average annual growth rate* through 1995–2005 of about 19 percent, doubling China's car population every four years. India is close behind.

The following is of course a fantasy projection, but its only proviso is that India and China will sustain the growth rates for car production and ownership that were experienced by the US, New Zealand, South Korea, Japan, Canada, Australia, France, Germany, Italy, Spain, and other "leading industrial nations" for a period of less than 20 years. If their growth rates are higher, the period needed to attain "saturation ownership" (the US's current rate) will of course be lower.

If China and India were to do this, the entire oil supply of the OPEC group, about 33 Mbd production and 25.5 Mbd exports, would not even satisfy these two countries' car fleet oil requirement! The fantasy Hydrogen Economy, and of course the biofuels, is therefore the only way out for current political and business leaderships – when or if they care to forecast any further out in time than perhaps the next five to ten years.

THE APOCALYPSE WAGON[9]

We assume that the fantasy fleets of China and India operate at much less than the average vehicle kilometrage per year of the EU6 (the core six nations), that is 22,000 km/vehicle/year.

The EU27, we can note, has a lower average annual kilometrage, but in all the European Union countries outside the "core group," average annual distances run are rising with economic growth.

Average fleet-wide car fuel consumption in Germany is 7.9 liters per 100 kilometers (L/100 km) and well above 12 L/100 km in the ultimate "role model" for motorization, the US. We could play "reasonable" and assume that the typical rate of annual oil demand for the Apocalypse Wagon is rapidly reduced in India and China, to two-thirds of the US benchmark, to about 8 L/100 km. We will also assume that India and Chinese cars will only travel 18,000 km/year. Annual oil burn per car is therefore around nine barrels/year.

Assuming that the Chinese and Indian fantasy fleet reaches 400 million units by 2025 (230 million in China and 170 million in India), and annual oil burn per vehicle is nine barrels (1,440 liters), daily average oil demand is as follows:

Chinese fleet of 2025: about 5.7 Mbd
Indian fleet of 2025: about 4.2 Mbd
Combined total: 9.9 Mbd

This relatively "low-case scenario" is above 40 percent of current total net exports of the OPEC group, but we can easily go further. If China attained ZPG (or zero population growth), at its current 1.25 billion, and India did the same (which is even less likely), but both countries attain the Motor Nirvana of US ownership rates (around 745 vehicles/1,000 population), their car fleets would number about 800 million vehicles (India) and 935 million vehicles (China). Even with a "technology optimist" scenario under which these fantasy fleets only consume about six barrels/year each, their total consumption explodes to about 10 billion barrels/year, or around one-third of the world's current total oil burn.

FIGHTING FOR FUEL

We therefore have a laughable fantasy, or an insight into exactly why three nuclear armed powers, China, India and the US, are ever more

The Apocalypse Wagon

Current world car fleet: about 800–825 million units
Current world land transport fleet: about 950 million units
(CEU or Car Equivalent Units including all motorcycles, scooters, buses, trucks, agricultural vehicles, construction + public works vehicles, etc)

Expressed in Car Equivalent Units, this 950 million CEU fleet consumes on average about 9 barrels oil or oil equivalent/CEU/year. Construction energy requirements are about 6–7 barrels oil equivalent/CEU, of which about 2.25 billion barrels are oil

Current US car fleet: about 200 million cars + 25 million CEUs
total: 225 million CEUs Oil burn: 9.5 million barrels/day

Figure 10.1 The Apocalypse Wagon

Source: Andrew McKillop, Juno M.E. Asset Management, LLC, New York.

likely to fight amongst themselves, or confront EU importers, which include two declared nuclear weapons states, for *the last oil reserves of the planet.* Under any hypothesis – excluding childish technological fantasies, and utopias – there is simply no prospect of China, India, or other countries including Malaysia, Slovenia, Brazil, Turkey, Iran, Ukraine, Mexico, the Czech Republic and other emerging car producers, being able to achieve US, West European, Australian, or Japanese rates of car production and ownership.

The Chinese Car Bomb therefore ticks onward, as each day another estimated 175,000 new cars and 57,000 motorcycles and scooters are produced. Each car or car equivalent (about three motorcycles or 0.25 heavy trucks) requires an average of about 2.5 barrels of oil and another 3.5 barrels equivalent of other energy inputs to produce, and must operate on bitumen-based highways, on tyres that themselves are about 40 percent oil by weight.

Not only is this explosion of the world car fleet a serious threat to the earth's environment, but its oil demand impact will become a threat to international peace and stability.

NOTES

1. Offroad and agricultural motor vehicles are in fact extensively, perhaps increasingly, used for human and goods transport on roads and tracks. This is notably the case in China and India, where the current lack of road motor vehicles incites the usage of agricultural vehicles for human transport (passengers riding in towed trailers). It can be noted that fuel efficiency of these "road vehicles" is very low because of technical reasons, that is, vehicles designed for slow speed offroad being used for road transport.
2. Recent Chinese car sales growth: "China's Car Sales Hit One Million for First Time," Reuters, December 16, 2002: "Shanghai: Annual car sales in China have topped the one million mark for the first time as a rising urban middle class crowns the world's fastest-growing market for foreign automakers, industry executives said on Monday. An official at the China Association of Automobile Manufacturers told Reuters 1.02 million cars were sold in China in the first 11 months of this year, representing a stunning 55.4 per cent jump from the same period in 2001."
3. Ross McCluney, "Population, Energy, and Economic Growth," Chapter 13, this volume.
4. "Remarks by Energy Secretary Spencer Abraham at the Global Forum on Personal Transportation," Dearborn, Michigan, November 12, 2002, from a US Department of Energy Press Release, reproduced at EROEI.com, www.eroei.com/articles/2004%11articles/energy-secretary-spencer-abraham/, accessed March 5, 2008.
5. M. Tadashi, *The Motorisation of Cargo Transport in Japan*, Hakuto-Shobu, Tokyo, 1982; Japanese Department of Transport for recent data; K.

Shimokawa, "Japan: The Late Starter Who Outpaced Her Rivals," in *The Economic and Social Effects of the Spread of Motor Vehicles*, ed. T. Barker, Macmillan, London, 1987; Japanese Department of Transport, *White Paper on Transport*, 1984.

6. G. Konn and Y. Okano, *Study of Modern Motor Transport*, Tokyo University Press, Tokyo, 1979.
7. Shimokawa, "Japan: The Late Starter Who Outpaced Her Rivals."
8. UK Society of Motor Manufacturers and Traders (SMMT), www.smmt.co.uk/home.cfm: statistics for 2001 showed the UK having 28.6 million cars and 3.5 million commercial vehicles in use. This compares with the UK 1951 car-ownership figure of 2.3 million, 1971 of 12.35 million, and 1981 of 15.63 million.
9. "Global Car Production Statistics Pages," www.geocities.com/MotorCity/Speedway/4939/carprod.html, accessed October 11, 2007. This site provides recent data on worldwide car production (from 1995). The very fast growth rates of car production in several countries (ten to fifteen times their population growth rates) is clear. This indicates total production in 2001 of about 40.9 million private cars, or about 112,190 per day, worldwide.

11

Venezuela, Chávez, and Latin American Oil on the World Stage

Sheila Newman[1]

> In 1998, or 168 years after independence, a tiny wealthy elite was separated by a vast chasm from the rest of the people, of whom one quarter were unemployed. This seems disgusting when you realize that Venezuela was then the second biggest [oil] exporter in the world and had received around 300 billion dollars in oil sales – or the equivalent of 20 Marshall Plans – over the preceding 25 years. It was in this context that Hugo Chávez and his social plan won the elections of 6 December 1998 with 56.24% of the votes.[2]

President Hugo Chávez is a social revolutionary with a giant budget. In 2006 the Energy Information Administration (EIA) ranked Venezuela ninth in world oil producers and sixth in world oil exporters.[3] For many of his countrymen Chávez appears to be seen as a towering figure of hope for rescue from a nightmare which began in 1498. But his Anglo critics portray him as an ogre treading clumsily over political alliances and destroying Venezuela's oil assets.[4]

Clearly Venezuela is pursuing a different political paradigm from that of the North American-led Anglophone countries. The Chávez government endorses a Christian socialist philosophy directly opposite to the Protestant capitalist one of a wealthy elite divinely elected on earth. In the Chávez philosophy, Christ was the first socialist, sharing wealth among the poor; a rich man might only enter heaven by giving away his possessions to the poor; a good leader should give everything to his country.[5]

One political explanation for this difference is that Latin America "missed out" on the progress model which dominates North America because, colonized by medieval Catholics, it was isolated from the development of Protestantism.[6]

Sociologist Max Weber theorized in *The Protestant Ethic and the Spirit of Capitalism* (1906) that Calvinism was the midwife of capitalism, delivering to the world the concepts of the "work ethic" and of election to earthly prosperity as a reflection of God's grace.

The work of Australian engineer and social analyst Sharon Beder supports a contrary view that the work ethic plus the progress model are driving the world over a cliff,[7] and this is pretty much Chávez's expressed view. Chávez also apparently shares a similar perspective to Al Gore's on global warming, but that is where the similarities end.

From the fifteenth century the indigenous long-term stable clan and tribal populations of Chávez's people were ravaged by invasion, immigration, disease, dispossession and slavery. The original peoples nearly died out, then, completely disorganized, ballooned in circumstances where child labor was the only source of additional income for low-wage landless people.[8] What is now called Venezuela contained a stable population estimated at around 400,000 Amerindians in 1498.[9] (Now the population is around 27 million.) In the early sixteenth century King Charles Martel V granted Welsers German banking firm rights to exploit the people and resources of Venezuela in payment of a debt. The colony returned to the Spanish Crown within 20 years and hereditary land grants were made to conquistadores for a time, but later declared illegal. Meanwhile the Amerindians fought back until smallpox overwhelmed most of them in 1580.[10] Not until 1821 did Simón Bolívar win the long indigenous struggle for independence.[11]

In 1921 the discovery of oil permitted agricultural and industrial development. At the start of World War II Venezuela's oil production was exceeded only by that of the United States. Much of the oil concession development involved attracted US, British, and Dutch companies. Venezuela became a democracy in 1958 and founded the Organization of Petroleum Exporting Countries (OPEC) in 1960.[12]

The historic inequities of colonial land distribution[13] guaranteed a large population of impoverished rural laborers. As oil prices waxed and waned, productive agricultural holdings were neglected and waves of poor people left the country regions to look for work in the city, creating the slum of Caracas. Between 1959 and 1964 the government redistributed rural land to 150,000 families, but many resold the land to speculators, it is said, because they had little education about farming and no ready market for their product.[14] Other wealth redistribution and educative policies were carried out but these programs failed to establish themselves against a background of depressed commodity prices and political schism. The then Democratic Action (DA) government was aligned with the US but many Venezuelans were sympathetic to the Castro regime in Cuba, which was charged with supplying arms to guerrillas in 1963.[15]

The state became increasingly repressive in the context of continued political unrest. In 1968 the Social Christians (SC) won government and remained in power until 1973.[16]

In the wave of nationalizations following the first oil shock, the DA government created the state-run oil and natural gas company, Petroleos de Venezuela SA (PdVSA), in 1975–76. PdVSA is Venezuela's largest employer and provides 80 percent of export earnings but, reflecting later trends to privatization, government revenue declined from 70.6 percent in 1981 to 38 percent in 2000.[17]

The oil countershock of 1979 culminated in currency devaluation by one-third and a change to an SC government, which remained in power until 1983, when DA was returned under Jaime Lusinchi. Despite promises to diversify the economy and deliver on housing, public health, and education, the situation continued to deteriorate. In 1988 another DA president, Carlos Andrés Pérez, introduced an austerity regime, removing subsidies on gasoline as well as on a number of important consumables, culminating in hunger riots in Caracas, with a death toll of thousands.

Two attempted military coups took place against a background of continued repression in 1992 and Hugo Chávez led one of them. President Pérez later went to prison for 28 months with the government limping along under another recycled leader, Caldera, whose foreign policy was very US-friendly. In 1995, 103 percent inflation hit the Venezuelan middle class. In 1997 doctors, university professors, and national telephone company workers went on strike. In December 1998 Hugo Chávez won the presidency.

On December 30, 1999, Venezuela's 26th constitution was approved by 71 percent of votes. The Senate was replaced by a single chamber National Assembly, and the Bolivarian Republic of Venezuela came into being, named after the national hero. Presidential terms increased from five to six years and limitations on Presidents serving a second consecutive term were lifted, but it became possible for the public to sack a President through a publicly initiated referendum. Privatization of the oil industry, social security, health care, and other major state-owned sectors was outlawed.[18]

According to the EIA, "Nearly one-half of PdVSA's employees walked off the job on December 2, 2002, in protest against the rule of President Chávez."[19] But another report says that they were prevented from working in a "bosses' lock-out" where "a small group of managers, directors, supervisors and technicians organised the sabotage of production and brought the industry almost to a halt," and Georgetown politics Professor Arturo Valenzuela commented that "The opposition ... has also been extremely irresponsible in trying to demand [Chávez's] resignation rather than trying to seek an electoral solution."[20] If we assume that PdVSA management was responsible for the declining returns to the state by PdVSA over the decades, then the view that this was a "lock-out" to preserve an undemocratic status quo by discrediting the Chávez government seems persuasive. Chávez had provoked US insecurity about oil supply by criticizing the Free Trade of the Americas Act (FTAA) and US foreign policy. The Chávez government had sacked some directors of PdVSA who were in political disagreement with the Venezuelan executive. These people then led calls for a general strike along with a variety of opposition parties and the Fedcamaras (Venezuelan Chamber of Commerce), who are supported by the US National Endowment for Democracy. A group marched on the presidential palace demanding Chávez's resignation, which the President refused. He was arrested and imprisoned. Pedro Carmona, President of the Venezuelan Chamber of Commerce, which receives funding from the US National Endowment for Democracy, was installed as Venezuelan President on April 11, 2002. On April 12, the US President's spokesman, Ari Fleischer, endorsed the Carmona government. But, on April 13, the presidential guard and the army arrested Carmona. Next the opposition collected signatures from 20 percent of the electorate required under Chávez's constitution to initiate a referendum to sack the President, but Chávez won the referendum.[21]

The distribution of PdVSA income had been increasingly diverted to private concerns, with returns to the state falling from 70.6 percent in 1981 to 38.6 percent in 2000. Despite permanent damage to production from sabotage in the industrial disputes of December 2002, Chávez's intervention had raised PdVSA returns to the state from 38.6 percent to 50 percent by 2004.[22]

Venezuela has for some time been a food importer, due to the country's very poor system of land management, which Chávez has begun to rectify in a major scheme. He seems to be seeking regional self-sufficiency, with protection for local production. He is opposed to overconsumption, openly warning about oil depletion. He is highly critical of US human rights abuses, at home and abroad, and opposes free marketism.

Obviously Chávez's regime threatens many established interests in a seething international struggle for resource hegemony. The economy is still in recession and maintenance and consolidation of the section of the population which supports Chávez will surely require that he carry out his promises. Perhaps Chávez's friendship with Castro will be a source of survival skills, and also his policy of strengthening regional Hispanic alliances.

There are a number of likely alliance candidates, including Mexico, which has begun to import food from the US under the American "free trade" agreement. Brazil, sensibly seeking independence from petroleum, was apparently counseled to drop its independence policies in exchange for leniency on international debt.[23]

Chávez actively seeks more diversification in petroleum trading, initiating a "South–South diplomacy" with sidelined and emerging polities in controversial and political oil trade accords with Cuba, Argentina, Uruguay, Brazil, Spain, Iran, Libya, Nigeria, Qatar, India, China, and Russia.[24] An agreement in October 2004 means that Russian oil sales to the US are actually honored by Venezuelan oil and Venezuelan sales to Europe are supplied by Russian oil. The Chávez government has paid off US$538 million of Argentinian debt and has agreed to provide contracts worth US$500 million to Argentina.[25]

In a fascinating avoidance of petrodollars, Chávez supplies 80,000 barrels of oil to Cuba a day, at a friendly price, with 20 percent of payment in the form of the supply of 150,000 Cuban doctors to the Venezuelan health service.[26]

Most importantly, Chávez hopes to create a Latin American petroleum company, "Petrosur," which would unite the public companies of Argentina, Bolivia, Brazil, Uruguay, Ecuador, and Venezuela.

Mostly Anglo analysts have intimated that Anglo oil companies won't touch Venezuela because of Chávez, and that the bituminous deposits of the Orinoco won't get developed, through lack of experts. But this is beginning to look like sour grapes as plenty of the non-Anglo oil companies – Russian LUKoil, China's CNPC, Indian ONGC, and Brazil's Petrobras – don't seem to be put off.

Writing for ASPO-USA, Dave Cohen reports: "Chávez gleefully announced that 'the United States as a power is on the way down, China is on the way up. China is the market of the future' after his meeting with CNPC President Jiang Jiemin." He concludes: "The bummer for the OECD nations is that El Presidente just might be right. The gold rush is on, but now excludes greater participation by Western international oil companies."[27]

Chávez has also not neglected regional diplomacy among the underworld of arms trade and revolutionary militia.[28] And, since an attempted putsch in 2002, Chávez relies on Cuban Intelligence for personal protection. Not surprisingly, the US government disapproves.

On December 4, 2006, Chávez won his third six-year term as President. In 2007 a referendum to make Chávez President for life was democratically defeated by 51 percent.[29]

In the light of Venezuelan social and economic performance in the decades preceding Chávez it would be hard for him to do worse than his predecessors, and he seems to be doing considerably better. Land redistribution is the basis of revolution and of social equity.[30] Venezuela recently signed an accord to give effective rights to its indigenous peoples.[31] Chávez has already begun allocating public land to the landless in a program accompanied by massive agricultural education. He has not as yet allocated any private land but there is the intention to repossess land which is being held for purely speculative purposes.[32]

And the Chávez government has better green credentials than any other petroleum producer. With an active commitment to mitigate the impacts of climate change and peak oil, it has initiated new public transport, has instituted organic farming as an important part of secondary school education, has facilitated a huge organic farm in the center of Caracas, and has plans for massive reforestation with the collection of 30 tonnes of seeds, and the planting of 100 million plants.[33]

The overwhelming positive signs of Chávez's example seem to be what we need for the twenty-first century. If this is really what Chávez is teaching and doing, then we can only hope that the whole world will unexpectedly come to its senses and follow.

NOTES

1. Although President Hugo Chavez's speech at the 60th General Assembly of the United Nations in New York on September 15, 2005 isn't specifically cited in this chapter, it was treated as very important in several of the documents referenced below.
2. Nicolas Lehoucq, "La redéfinition du rôle géopolitique Vénézuelien," Institut d'étude des Relations Internationales Paris, www.memoireonline.com/11/06/287/m_redefinition-role-geopolitique-venezuelien0.html, accessed September 21, 2007. (Translated here by S.M. Newman.)

3. (All hydrocarbon liquids) www.eia.doe.gov/emeu/cabs/topworldtables1_2.htm.
4. For instance, *The Economist* unreasonably ignores the poverty of Venezuelan people under the old management of oil in a one-sided interpretation of events in "Oil's Dark Secret," Economist Intelligence Unit, Executive Briefing, August 11, 2006, www.economist.com/business/PrinterFriendly.cfm?story_id=7270301.
5. Lehoucq, "La redéfinition du rôle géopolitique Vénézuelien."
6. In *The Pan-American Dream*, US conservative writer Lawrence Harrison attempts to explain the differences in economy, government, human rights, and standard of living in American Hispanic societies according to Weberian theory.
7. Articles by Sharon Beder on the Work Ethic. http://homepage.mac.com/herinst/sbeder/home.html#work.
8. UNICEF estimated that 9.9 percent of children aged 5–14 years in Venezuela were working in the year 2000. Government of Venezuela, *Multiple Indicator Cluster Survey (MICS): Standard Tables for Venezuela and Annex I: Indicators for Monitoring Progress at End-Decade*, UNICEF, 2000, www.childinfo.org/MICS2/newreports/venezuela/venezuela.htm, and www.childinfo.org/MICS2/EDind/exdanx1.pdf.

   Doepke hypothesized that fertility falls where policies such as education subsidies and restrictions on child labor affect the opportunity cost of education. The populations of South Korea and Brazil had begun to grow rapidly around the same time, but South Korea had an effective public education system, and strongly enforced child-labor restrictions, whereas Brazil had a weak public education system and poorly enforced anti-child-labor laws. M. Doepke, *Growth and Fertility in the Long Run*, Mimeo, University of Chicago, 2000, available in reduced form in M. Doepke, "Accounting for Fertility Decline During the Transition to Growth," *Journal of Economic Growth*, vol. 9, no. 3 (September 2004), pp. 347–83.
9. www.nationsencyclopedia.com/Americas/Venezuela-HISTORY.html.
10. Ibid.
11. Ibid.
12. http://news.bbc.co.uk/2/hi/business/2549589.stm.
13. www.forestpeoples.org/documents/conservation/Ven10c_jan04_ch6_eng.pdf.
14. Seth DeLong, "Venezuela's Agrarian Land Reform: More like Lincoln than Lenin," COHA, February 25, 2005, http://www.venezuelanalysis.com/analysis/963.
15. www.nationsencyclopedia.com/Americas/Venezuela-HISTORY.html.
16. Ibid.
17. Ibid.; Lehoucq, "La redéfinition du rôle géopolitique Vénézuelien"; Arnaud Doizy, "La politique étrangère des Etats-Unis au Venezuela, la période Chavez (1999–2007)," Université Panthéon-Assas Paris II, www.memoireonline.com/06/07/491/m_politique-etrangere-etats-unis-venezuela-periode-chavez2.html, accessed September 22, 2007.
18. www.nationsencyclopedia.com/Americas/Venezuela-HISTORY.html.
19. EIA, www.eia.doe.gov/emeu/cabs/Venezuela/Oil.html. See also, for other anti-Chávez sources, http://marketplace.publicradio.org/display/

web/2007/09/13/venezuelan_oil_expats_resurfacing/. On the US stance, Arturo Valenzuela, Professor of Politics at Georgetown University, commented on the PBS *Jim Lehrer Newshour* ("Troubled Nation," December 17, 2002), that "Unfortunately, the radicals on both sides are maintaining this conflict. The opposition, for example, in my view, has also been extremely irresponsible in trying to demand his resignation rather than trying to seek an electoral solution. In fact, the constitution as I said earlier does make it possible for Chávez to be submitted to a referendum in August of next year. It seems unreasonable not to focus on that. Chávez has said he would accept that as a possible outcome. The problem is that the opposition wants him out now. Chávez says I don't want to leave and the situation is getting worse day by day." www.pbs.org/newshour/bb/latin_america/july-dec02/venezuela_12-17.htm.

20. Jorge Martin, "Venezuela: Opposition 'Strike' or Bosses' Lock Out? An Eyewitness Account," www.marxist.com/Latinam/venezuela_eyewitness0103.html. See also www.thirdworldtraveler.com/South_America/Venez_Coup_Countercoup.html.
21. Arnaud, "La politique étrangère des Etats-Unis au Venezuela, la période Chavez (1999–2007)."
22. Lehoucq, "La redéfinition du rôle géopolitique Vénézuelien."
23. It relies on uncompressed gas but also, unfortunately, on biofuels which will lead to tragic soil and forest destruction. Nicolas Lehoucq writes that Brazil had achieved a "quasi-independence from petroleum" in the 1990s and has even developed cars which the driver can select to function, by a simple switch, from gasoline to non-liquefied gas to ethanol. "But this innovatory system was threatened by the World Bank. President Lula was attempting to obtain a partial cancellation of Brazil's debt and the World Bank attempted to negotiate a deal whereby Brazil would cease its petroleum independence program." Lehoucq remarks that the World Bank is US dominated and wonders if the US uses petroleum as a means of control of third world countries. Lehoucq, "La redéfinition du rôle géopolitique Vénézuelien."
24. Ibid.
25. "Using Oil to Spread Revolution," *The Economist*, July 28, 2005, Caracas, cited in Informaciones tomadas de Encuentro en la Red Informaciones de Cubanet, www.cubadata.com/chronology/2005/8-August%202005%20All%20News.pdf, accessed March 5, 2007.
26. Lehoucq, "La redéfinition du rôle géopolitique Vénézuelien."
27. D. Cohen, "Venezuela – Aló Presidente!" ASPO-USA publication, August 29, 2007, www.aspo-usa.com/index.php?option=com_content&task=view&id=202&Itemid=91.
28. Lehoucq, "La redéfinition du rôle géopolitique Vénézuelien."
29. "Hugo Chavez's Annus Horribilis," M&C News, Americas, December 14, 2007, http://news.monstersandcritics.com/americas/news/article_1380925.php/YEARENDER_Hugo_Chavezs_annus_horribilis.
30. A fact strangely overlooked in many studies of the French Revolution, where analysis of methods undertaken for the redistribution of land affords a remarkable perspective.

31. As with Australia's *terra nullius*, no treaty had ever been undertaken with the indigenous people. Source: The Forest Peoples' Program, "Protecting and Encouraging Customary Use of Biological Resources: The Upper Caura, Venezuela," www.forestpeoples.org/documents/conservation/Ven10c_jan04_ch6_eng.pdf.
32. DeLong, "Venezuela's Agrarian Land Reform: More like Lincoln than Lenin."
33. Derek Wall, "Viva Venezuela verde!" http://commentisfree.guardian.co.uk/derek_wall/2007/04/viva_verde_venezuela.html; and Eva Golinger, "Venezuela's Green Agenda: Chavez Should be Named the 'Environmental President'," February 27, 2007, Venezuelanalysis.com, www.venezuelanalysis.com/analysis/2244, accessed September 22, 2007.

Part III: The Big Picture – False Solutions, Hopes, and Fears

> Without cheap fertilisers – and the cheap oil used to make them – this productivity can't be sustained. As oil prices continue climbing this century, this cycle may stall with disastrous consequences. We burned more than a trillion barrels of oil over the past two decades. That's eighty million barrels a day – enough to stack to the moon and back two thousand times.[1]

This part discusses climate change mitigation, resource limits including soil and biofuels, population numbers, and poverty, in the global context. It looks hard at four proposed big technology solutions – cellulosic biofuels, nuclear fission, nuclear fusion, and deep geothermal, including "hot rocks."

Andrew McKillop's "No Choice but International Energy Transition" follows on from "Oh Kyoto!" in the first edition of this volume, as he describes how much more needs to be done to mitigate climate-change catastrophe and an oil and gas cliff. He combines insider knowledge of energy sector administration in the Organization of Arab Petroleum Exporting Countries' (OAPEC's) technology transfer subsidiary AREC, and international energy policy making gained as Policy and Programming adviser to the European Commission's DGXVII – Energy, with recent finance sector experience in New York, to interpret the labyrinth of corporate interests and political decision-making naivety. He also flags the danger that the destruction of natural habitat to grow "biofuels" could be used by big business to ensure business as usual by writing biofuels off as "carbon credits."

Updated from the first edition, "Population, Energy, and Economic Growth," by Ross McCluney, reviews both reality and moral conflicts inherent in energy limits, population growth, and economic growth. His "Renewable Energy Limits" (substantially rewritten from the first edition) provides an excellent overview and basis from which to consider some of the other chapters in this part which focus on particular energy sources.

Taking up the issue of biofuels in general and cellulosic biofuels in particular, Alice Friedemann's "Peak Soil" reminds us, in her evocation of its microscopically organized biodiverse systems, that the soil we are exploiting to death is alive. A few years of trendy marketing could spell planetary desertification by commercial destruction of our most basic resource.

I wrote the chapter "Nuclear Fission Power Options" myself for a number of reasons. It was very difficult to find an author who was not employed in the area and could be seen to speak freely or whose treatment of various nuclear technologies was not affected by an initial pro or contra nuclear position. I designed the chapter to cover, in as few words as possible, what I could myself glean from recent material and then sought answers for questions that arose from this material. It seemed to me that many readers might appreciate a more basic introduction to the phenomenon of nuclear power than is usually available before authors launch into its technology. It also seemed necessary to cover thorium as a fuel and the prospects for fast breeder reactors using thorium or other fuels. It became apparent to me that politico-economic and investment concerns affect the mediatization and promotion of nuclear fission power at this time as much as hard science. Reader awareness needs to go beyond the immediate costs of projects or the traditionally accepted dangers of nuclear to the commercial politics and interests at national and global level. My father, John Newman, who is a geophysicist with experience in uranium exploration in Australia, did a lot of the background research into this, which I interpreted sociologically. The chapter was peer-reviewed.

Michael Dittmar's chapter, "Fusion Illusions," about the viability of commercial fusion, fills a huge gap in general knowledge and will be a valuable addition to many university courses. Not only does it give answers to questions about fusion, but it frames the questions most of us would not have thought to ask. As a particle physicist, Dittmar gives us a practical but sub-atomic view of the issues arising in a field made up of exotic materials like deuterium and tritium interacting with lithium "carpets." Whilst the reader may not become fluent in sub-atomic behavior just by reading this work, most will emerge with the satisfying feeling of having a basic ability to discuss what is meant by commercial fusion and whether it is likely to be achieved.

Deep geothermal energy is a very exciting but still in-progress area of research where it is hard to find articles, so my chapter, "Geothermal" resumes and comments on what information is around.

"Notes on Terra Preta": in this part I would have liked to have included a comprehensive chapter about research into the old soil technology known as "terra preta" and "agrichar," which may prove in the long run to be of more importance to the world than any other proposed solution. Unfortunately there was only the time and space to include a summary of the subject.

NOTE

1. David R. Montgomery, *Dirt*, University of California Press, Berkeley and Los Angeles, 2007.

12
No Choice but International Energy Transition

Andrew McKillop

INTRODUCTION

No "supply side solutions" have yet been able to replace or substitute for current rates of world oil and gas consumption, let alone for growth in demand for these declining fossil fuels. The "decline denial industry" has emerged to argue that there are. Underpinning this "denial industry" is the knowledge that vast consumption of fossil fuels is needed for maintaining economic growth, plus more hidden, anguished reasons such as "defending our civilization."

At the same time, the climate change gas emissions due to burning these fuels at a yearly rate of about 6.7 billion tonnes oil equivalent (for oil and gas) are accepted as alarming, needing change, and concerning everybody.

The Middle East, which holds about 60 percent of the world's remaining oil and about 50 percent of remaining gas reserves, is already the theater of civil and international war, driven by "Great Power rivalry" for control over diminishing fossil fuel reserves. In the Middle East and Central Asia, conflict over oil and gas and the struggle for control over reserves or transport routes can at any time become all-out regional war, permanently damaging energy production, export installations, and devastating local environments.

This situation is accepted, by many, as a sort of fatality and therefore "normal."

Climate change mitigation attempts and plans, as well as the Kyoto Treaty, now include many proposed "energy protocols" and "carbon-related" initiatives. Unfortunately the key requirement of oil and gas intensity reduction – doing more with less, or less with less, but in any case using less – is totally absent from the scene. For the Organization for Economic Cooperation and Development (OECD) or "rich world" countries, the need to reduce average per capita oil and gas consumption is urgent. It must be implemented now and applied over the long term for any chance of energy transition without economic crises and social conflicts inside the OECD countries. Some emerging industrial countries, especially China and India, are unable to reduce the growth of their oil and gas intensity due to very fast industrial growth and urban development, exactly like that of the OECD countries in the period of postwar reconstruction, about 1950–73.

Demand and consumption of oil, gas, and coal measured about 5.8 billion tonnes/year in 2006 and were growing fast. All these fossil fuels have their depletion curves, with those of oil and gas closest. All are vectors of runaway climate change. The first part of energy transition requires acceptance of these facts and consequent reduction in the use of these fuels.

Because we are left with the fossil fuel coal, and uranium, as relatively abundant fuels, and the renewable energy sources, the development of renewables will be another part of energy transition. Coordinating and funding this worldwide and long-term effort is unlikely to be possible if it relies solely on "market mechanisms." An international and multilateral effort for energy transition is necessary, removing oil and gas from the supercharged arena of market trading, automatically funding short- and long-term development of renewables on a worldwide basis, restructuring the economy and society of both OECD and non-OECD countries for reduced energy intensity.

THE LURE OF THE SUPPLY SIDE

It is no coincidence that the so-called New Economy and its "neoliberal" ideology is based on and calls for "supply side answers" to any economic problem. When these do not work the New Economy answer is "demand destruction," after which reduced supply is supposed to satisfy the reduced demand. In fact this only transfers demand to the future.

The best example of this is Russia's economic collapse in the 1991–96 period, during which "demand destruction" for oil and gas was very intense, with about a 45 percent cut in Russian domestic consumption. This of course permitted large increases of Russian oil and gas exports, reducing world prices and maintaining an appearance of abundance.

Today, after more than six years of strong economic growth, growing Russian domestic gas and oil demand, and diminishing oil production, have already led to declining net oil exports from

Russia. Declining gas exports may soon follow, perhaps as soon as 2009. In other words, "demand destruction" in Russia was temporary because no restructuring of the economy or society took place. With a return to economic growth and reduced poverty, domestic energy consumption increased annually at about 4.5–6 percent for oil and gas.

Russia's temporary decline – economic, geopolitical, and of its domestic energy consumption – was treated as "long term" by adepts and defenders of the New Economy. In fact the Russian case was perhaps unique, and is neither a model for us all nor a long-term model for sustainable adjustment to declining oil and gas reserves worldwide.

Energy transition requires restructuring and deep modification of the demand side to permanently reduce oil and gas intensity in the sure perspective of diminishing world production and supply. In turn it will become necessary to change today's energy and economic infrastructures, which go hand in hand with constantly expanding electrification. Supply side myths, such as the imagined abundance of cheap uranium and reliability of nuclear power which lead to a belief in electricity as a solution to diminishing fossil fuels and runaway climate change, must be smashed. World electricity demand in 2006 is growing at about 9 percent per annum, and is about 79 percent generated from oil, gas, and coal.

The energy-dependent economy, in constant growth and change, receives and generates successive overlays of new fossil energy dependent consumer technology, such as the so-called "communications revolution," feeding back to continued growth of energy demand. This produces an apparent imperative of finding supply side solutions, which are being set as the only way to avoid chaos and conflict. In fact the exact opposite applies.

The cult of economic growth, consumer technology innovation, and the confused and impossible desire for "universal prosperity" are collectively taken to be "progress." Anything going the opposite way is taken to be retrogression and failure. This may well be how supply side solutions to the "energy crisis" are thought about in the fossil energy intensive societies, but on the ground, in the Middle East and Central Asia, the hunt for cheap oil and gas results in military invasion, civil war, and devastation. This barbarous "strategy" of military domination to ensure future supplies of oil and gas is the ultimate "supply side policy," and events in Iraq and Afghanistan, Lebanon, and Palestine show that it is a failure. Impacts on world climate from burning a total of about 11 billion tonnes/year of fossil fuels are easy to see. The future damage by runaway climate change to human society – including the economy – is increasingly easy to forecast.

TARGETS AND MEASURES FOR ENERGY TRANSITION

We need to move a lot further and faster than the Kyoto Treaty to reduce oil and gas intensity in the OECD countries, to limit the growth of fossil energy intensity in the emerging non-OECD industrial countries, and to develop the renewable energy sources on a worldwide, coordinated, and automatically funded basis.

ENERGY INTENSITY REDUCTION IN OECD COUNTRIES

The urgent need for reducing oil and gas intensity of the OECD "rich world," or mature urban "postindustrial" societies is at present little recognized, or at least rarely admitted. Reasons for this have been discussed above. The present situation regarding average oil and gas demand per capita, in barrels or barrels oil equivalent per year, is shown in Table 12.1.

Table 12.1 Oil and gas intensity

Country or region Rounded averages, 2006	*Oil intensity (barrels per capita per year, bcy)*	*Gas intensity (barrels oil equivalent per capita per year, boecy)*
United States	25.5	13.4
South Korea	17.5	5
Japan	14	5
Italy	12.4	8.2
Germany	11.8	6.9
Turkey	4.5	4.5
China	2.5	0.3
India	1.3	0.25
Low-income Africa	0.5	0
World Average	4.8	2.75

Sources: Based on various data sources including UN agencies, *BP Statistical Review of World Energy*, IEA, and others.

Electric power demand intensities, we can note, follow about the same distribution but with much higher variations between extremes.

Low-income African and Asian countries consume around 100–150 times less electricity per capita than high-income OECD countries, rather than "only" about 20–50 times less oil and gas per capita. It also seems that absence of gas pipeline transport infrastructures serving South Korea and Japan (dependence on liquefied natural gas, LNG) has reduced growth of their gas intensities relative to other OECD countries with gas pipeline infrastructures. Importantly, we can note China and India's low relative oil and gas intensities. As we know, their energy demand growth is high and sustained, reflecting rapid industrialization and high economic growth rates.

This is the current situation, and we can, from this, immediately set a rational target for oil and gas intensity reduction in the OECD countries and attain the 2006 world average within ten to fifteen years, sooner for oil and later for gas, depending on national, regional, and international conditions. These conditions and factors will concern gas intensity reduction for the US and European countries, which depend on surely diminishing gas supplies from single-country suppliers already facing gas production difficulties; Canada in the case of the US, and Russia in the case of EU countries. The Kyoto Treaty, anti-oil, pro-gas, and pro-electricity in its real world impacts, obliquely recognizes that oil intensity reduction will be the most urgent task.

OIL AND GAS INTENSITY REDUCTION IN OECD COUNTRIES

Necessary, rapid, and severe reduction of oil and gas reliance in the light of their depletion profiles and their role in climate change requires planning. The impossibility of achieving this through "market only" mechanisms can be gauged from the exceptionality of any decline in world oil demand. In the entire 60-year period from 1945 to 2005, world oil demand declined long term only once – for each of the first three years of the very intense worldwide economic recession of 1979–83. Significantly, this decline was not accompanied by a decline in world natural gas consumption.

This period of severe economic hardship was marked by destruction of the economic tissue, and loss of economic and social infrastructures. Ten or fifteen years of such decline would be intolerable and hard to imagine. Further, the rate of decline or compression of oil and gas consumption would need to be up to 7.5–9 percent per year depending on country, sustained on a year-in, year-out basis. This is far higher than the actual 3.5 percent per year average decline, only in world oil demand, that was achieved in 1980–82.

It will take shocks to wake up decision makers and the public to the stark fact that peak oil and peak gas are not debating chamber, academic subjects, but concern everybody. After that, proposing and selecting plans for long-term compression of oil and gas consumption must follow. The time taken for this will have to be much less than the "Kyoto process," which took about 13 years, from 1992 to 2005, to move from the stage of international debate to on-the-ground action in the ratifying countries, albeit only in the shape and form of the carbon finance and carbon credits circus.

This circus is really a blank check for building gas-fired electric power stations, and for raising the price of agrocommodities through plans for diverting large amounts of basic agricultural products to biofuels production. As a way for achieving the stated goals of cutting greenhouse gas emissions in the ratifying countries through 2008–12, it leaves a lot to be desired.

International energy transition will need to start with the aim of reducing oil and gas intensity in the energy intense economies and societies, mostly the OECD countries. At the same time, very large-scale measures for rapid but coordinated development of renewable energy on a worldwide basis will be required. The likely "window of opportunity" for debating, discussing, negotiating, and putting the needed measures into action will unfortunately be short: peak oil is already here, and peak gas for EU countries dependent on Russian gas exports will likely start by 2009.

At present, only the wide variety of small-scale and uncoordinated measures favored by Kyoto Treaty interpretation (often a very liberal interpretation!) operate. As noted, these consist essentially in switching financial resources to gas-fired electric power generation, expanding wind power generation, bringing in some habitat and building sector measures for energy conservation and substitution of oil-based products, targeting biofuels production based on energy intense agroindustrial operations, and "calling for" nuclear power. Overall, it is not unlikely that "perverse impacts" of these measures, in general terms, will lead to increased oil and gas demand.

As intimated above, presently interpreted, the Kyoto Treaty amounts to a charter for building gas-fired power stations in ratifying countries. EU countries, according to the European Commission "Green Book" report of 2006, will construct about 28,000 MW of new gas-fired power plants in 2007–15. Kyoto Treaty interpretation does seek to "drive oil out of the energy mix," but coherent published targets and measures for achieving this are rare. One counter-example is Sweden, claimed by its political leaders to be embarked on "zero oil by 2020," at a moment in time when its Scania and Volvo heavy truck construction and exports are at record high levels. In other words, Sweden is very happy for its overseas customers to keep burning gas oil in heavy trucks, but wants to be "clean at home"!

Our coherent and target-based plan would set country targets for oil and gas intensity reduction, perhaps in five-year slices, initially with high annual reduction targets, modifiable in the case of persistent difficulties. This introduces at least two major features of our international energy transition plan that are almost absent from the Kyoto Treaty application and compliance procedure, and the majority of so-called "carbon protocols" aired from time to time.

OIL AND GAS SUPPLIES AND PRICES

We can be certain that financial markets have already "integrated" the approach of peak oil through 2003–06 by an "exuberant round" of massive price rises for primary products right across the board, from metals and minerals to soft commodities. These now receive the biofuels boost or price "premium" where they could or might be utilized in producing biodiesel or bioethanol. The driver for this is oil prices and energy prices. Oil prices are now both high and extremely volatile, while gas prices are low and extremely volatile. This is a trader's delight but useless for planning and achieving energy transition in an orderly framework.

Simply because market traders and operators considered oil prices "too high" and "unsustainable" after the summer peak of world oil demand in August 2006, oil prices were clipped by about 25 percent in a few weeks, only to increase again with the onset of cold weather in big consumer markets. On the supply side, world oil production and supply is now rigid, stagnant, and inflexible. It is unlinked and disconnected from the demand side. On that side "robust growth" is the reality, albeit denied or countered with misleading data that present world oil demand growth as at "very low rates". In addition, because oil prices were considered "too high" but not amenable to speculative downsizing in summer 2006, this tack shifted to world traded gas, resulting in absurdly low gas prices equivalent in energy terms to oil at about US$17 per barrel.

Apart from blurring public and political understanding of peak oil and peak gas reality and generating huge trading gains for successful market players, this incoherent market-based response to long-term decline in oil and gas supplies, and the urgent need to reduce energy intensity and develop renewables, is totally ineffective and unrelated to real needs. It is even counterproductive in the sense that it gives the impression that "if it's traded it has to exist." This again, however, is a distraction because at any one time the paper contract volumes of traded oil exchanged on world oil markets can be tens or hundreds of times actual world daily oil demand, about 86.5 million barrels/day on average in 2006, using an "all liquids" base. The oil exists on paper, but not in the production sites, pipelines, and storage tanks. The same applies to a lesser extent but with increasing frequency to world traded natural gas.

The solution is simple and radical: removal of oil and gas from the trading arena. This would require the creation of an international agency, modeled on the International Energy Agency, but including all consumers and major oil and gas producers. It could be called the International Oil & Gas Agency (IOGA), and be charged with deciding and allocating oil and gas supply volumes and prices on a 90-day forward basis. Progressive price increases for internationally supplied (cross-border) oil and gas would be applied by the IOGA which would issue regular notes and information concerning mid-term and long-term oil and gas reserves and production trends.

INSTITUTIONAL LINKAGE

The IOGA would respond to and interpret decisions made by the International Energy Transition Agency (IETA) which would necessarily be a UN-related agency similar to the coming UN Climate Change Agency or to existing UN-related and international institutions and agencies such as the OMC or MARPOL. To achieve coordinated, internationally funded development of renewable energy sources worldwide, and energy economy and economic restructuring for reduced oil and gas intensity in the OECD countries, would require the creation of an energy finance institution. This could be modeled on the International Monetary Fund (IMF) and be called the International Energy Fund (IEF).

Like the IMF, this Fund would act to ensure compliance with oil and gas intensity reduction measures in applicable countries (mostly OECD), ensuring automatic and sufficient funding of worldwide projects for large-scale development of renewable energy. Similar to the "Special Drawing Rights" (SDRs) system of IMF loans in economic emergency conditions and to the reciprocal obligations of drawing countries to carry out "good housekeeping" measures (conditionality of loans), the IEF would see that countries complied with set and agreed national oil and gas intensity reduction plans and targets. Such international Oil Drawing Rights (ODRs) and Gas Drawing Rights (GDRs) could be modeled on the SDR system. These financial instruments would be applied to member nations in the compliance system for oil and gas intensity reduction.

The Kyoto Treaty system and framework of obligations with its rather incoherent and ineffective system of financial "sticks and carrots" could be absorbed into the new international energy transition plan and process.

International energy transition measures must include accelerated and worldwide development of the renewable energy sources. The IETA would set world, regional and national targets for renewable energy development, and the IEF would provide adequate, automatic finance for these targets. An operating and technical agency will also be required, the International Renewable Energy Agency (IREA). This would integrate UN agencies such as the UNDP and UNCNRET, but would also require very large and permanent technical resources and manpower. Very close linkage between the IEF and IREA would be needed for the lifetime of these institutions, which would need to be at least 50 years.

NEAR-TERM REALITIES

It is thus possible to sketch out potential or possible international agencies, structures, measures, and arrangements for achieving energy transition. However, this will not happen though without public and political demand for transition. Demand presupposes acceptance of need for transition. Such public acceptance is frankly unlikely, because of the many social, economic, political, and cultural factors that support and maintain current profligate and unsustainable fossil energy consumption in the "rich world," and very fast growth of fossil energy consumption in the emerging economies.

The public acceptance in many countries of an urgent need to mitigate then reduce climate change should not be confused with action. Public and political concern in countries involved in "oil war" or "resource war" in the Middle East and Central Asia does not extend to doing anything about the real world causes of their country's involvement in these wars. Their involvement arises from their intense dependency on imported oil and gas. Public interest and concern about so-called global terrorism generates a steady stream of pulp paperbacks, TV docu-dramas and panel discussions on al Qaeda and the "Clash of Civilizations," but the "Terror of the 100-dollar Barrel" remains the real stumbling block to moving forward to energy transition.

In the near term the arrival of peak oil followed quickly by peak gas (perhaps as early as 2009 in the Euro-Asian region) will break down remaining barriers to public and political understanding of the linkages between the apparently disparate, but in fact interdependent "threats to civilization," and the need for change. Signs of this will be runaway price rises for oil, and increasing prices for traded pipeline gas in Europe and international liquefied natural gas (LNG) supplies. Extreme volatility will accompany the oil and gas trading "system" for as long as it exists. The media will of course maintain the idea of the "terror premium" or technical and industrial difficulties as the cause of continuing price rises for traded oil and gas, but this fig-leaf will quickly fall away. By the time this volume is published, we can suggest, this proposal for cohesive international measures and adequately skilled, staffed, and financed institutions to deal responsibly with oil and gas decline will no longer be casually dismissed.

13
Population, Energy, and Economic Growth: the Moral Dilemma

Ross McCluney

Before the fossil fuel-driven industrial revolution and the improved transport of foodstuffs, goods, and persons that this enabled, population density correlated with resource availability. A too-large population could not long live beyond the limits of the local ecosystem's carrying capacity. As transportation systems advanced due to the ever-increasing availability of cheap fossil fuel-based energy, physical and biological carrying capacity limits could be exceeded in a region by importing resources from other regions. This can continue for as long as energy remains cheap and abundant, and while environmental and economic impacts remain tolerable for human populations.

The process has now been carried to extremes; our world's human population exceeds the physical and biological carrying capacity of the whole earth, made possible solely by fossil fuels. Wackernagel and Rees developed the concept of an *ecological footprint*, the biologically productive land or sea area required to produce sufficient resource yields for the supported human population, and to absorb the corresponding carbon dioxide emissions.[1] The same method is presented in the US National Academy of Sciences publication, "Tracking the Ecological Overshoot of the Human Economy."[2]

Redefining Progress produced a November 2002 report outlining the ecological footprint of 146 nations. As Mathis Wackernagel, the Sustainability Research Program Director, notes: "Humanity's ecological footprint exceeds the Earth's biological capacity by about 20 percent," continuing: "many nations, including the United States, are running even larger ecological deficits. As a consequence of this overuse, the human economy is liquidating the Earth's natural capital."[3]

About three-quarters of the world's current consumption of resources is by the approximately 1.2 billion people living in what are called "rich nations," while the remaining quarter is consumed by the other 5 billion people currently living on this planet – almost a third of whom are categorized (by various United Nations (UN) agencies and other organizations) as living in great or extreme poverty. It would require at least three times the earth's entire resources and physical area to provide all the world's current population with the material and energy currently consumed by an average North American citizen. This immediately leads to questions not only on the physical possibility, but also the logical validity of economic expansion as a remedy for poverty. Since oil and natural gas drive all types of classic or conventional economic expansion, the peaking and subsequent decline of world oil production poses immediate and direct challenges to attempts at achieving unlimited global economic growth.

In 1997 geological consultant Walter Youngquist wrote:

> That oil production will peak and then decline is not debatable. If the more optimistic are right, and the peak date is a little further away than most geologists now predict, this would simply exacerbate our problems, for it means that the population at the turning point of oil production will be even larger than it would be at an earlier date, and it will then be more difficult to make the adjustment toward life without oil. Envisioning what the post-petroleum paradigm will be like involves consideration of myriad facets of the world scene. The worldwide decline of oil production, ultimately to the point where it is insignificant relative to demand, will have many ramifications, changing world economies, social structures, and individual lifestyles.[4]

The steady decline of world oil production after peak oil and over the next few decades makes for a sombre scenario regarding world food production and food availability per head of population. A significant decline in world population, due to this single factor, is more than possible. How we adjust as a global community to this challenge can only be of great concern to us all.

Proponents of staying on course – that is, "business as usual" – say that as we run out of oil it will be replaced by extreme energy conservation and a radical switch to renewable energy sources. At the same time, political leaderships resist any significant increase in oil and energy prices, while at every moment declaring their faith in "market mechanisms." Though technology improvements – but *not* breakthroughs – are possible and likely, few signs exist today that intensive energy conservation and a complete switch to renewable sources of energy is taking place, or is even being coherently promoted by world leaders.

The growing availability of cheap oil over the last century and a half has led to an enormous expansion of human population, industrial and technological impacts on the environment, and extreme dependence in the urban industrialized nations on cheap fossil energy. So long as cheap energy subsidizes and enables bulk transport of vital raw materials, food commodities, energy minerals, and industrial goods, and as fossil fuels decline and energy becomes inexorably more costly, humanity will reach a turning point. The current course cannot be continued indefinitely, and by this I mean *for more than a decade.*

RESOURCES AND POPULATION

Resources and goods are traded globally. A result is that regions with inadequate supplies of any input can make up for it with resources imported from elsewhere, as long as their demand is solvent (that is, if they have the cash for importing resources). If the raw materials for a factory are not locally available, they can be and are shipped around the world to where they are needed. If the oil products, natural gas, or electricity needed to run large transportation systems are not locally available, large pipelines, oil and gas tankers, and electric power grids will transport the energy to where it is used, and over great distances. If the soil is too poor, and fresh water inadequate to grow crops well in a region, fossil fuel-derived or powered machines, fertilizers, pesticides, soil conditioners, irrigation pumps, and so on, will be used to maintain output, or even temporarily increase agricultural productivity. Increased food production usually supports larger populations, and these populations depend upon imported, external resource inputs, fueled by cheap oil and other fossil fuels to maintain themselves.

The most important resource and single biggest item of world trade in volume terms in the so-called global market society is petroleum. Table 13.1 gives 2004 imports and exports of crude petroleum for selected countries, along with their populations, ordered by population size. The largest *importer* in the world was the United States, second was the relatively small Japan, having just under half the imports of the US. The combined population of the US and Japan was 427 million. The largest *exporter* is Saudi Arabia, with only 24 million people, but projected to increase to 35.6 million by 2025. Combining Saudi Arabia, United Arab Emirates, Iran, and Iraq, their total oil exports do not quite meet the imports of the US and Japan, and they have only 32.66 percent of the importers' population.

Table 13.1 World petroleum supply and disposition, 2004 (crude oil only)

Country	*Population in millions (data date 2006)*	*Primary supply crude oil imports (1,000 barrels/day)*	*Crude oil exports (1,000 barrels/day)*	*Net crude oil exports over imports*
China	1,311	2,449	110	–2,339
India	1,122	1,912	0	–1,912
US	299	10,088	27	–10,061
Brazil	187	450	230	–219
Russia	142	84	5,211	5,127
Japan	128	4,049	0	–4,049
Mexico	108	0	2,118	0
Philippines	86	200	0	–200
Germany	82	2,218	22	–2,196
Egypt	75	0	38	38
Iran	70	0	2,556	2,555
Thailand	65	870	70	–800
UK	61	1,124	1,223	99
France	60	1,718	0	–1,718
Italy	58	1,762	2	–1,759
South Korea	49	2,271	0	–2,271
Colombia	49	4	213	210
Spain	46	1,197	0	1,197
Algeria	38	7	1,279	1,272
Canada	33	923	1,336	414
Iraq	30	0	1,600	1,600
Saudi Arabia	24	0	7,143	7,143
Taiwan	23	1,004	0	–1004
Netherlands	16	1,039	24	–1015
Chile	16	208	0	–208
Angola	16	0	1,011	1,011
Ecuador	13	0	379	379
Hungary	11	109	3	–106
Sweden	10	415	0	–415
Israel	9	210	0	–210
Libya	7	0	1,219	1,212
Finland	5	216	0	–216
Singapore	5	881	0	–881
UAE	5	0	2,172	2,172
Kuwait	2	0	1,479	1,479

Source: Energy Information Administration, US. Department of Energy, www.eia.doe.gov/iea/, and Population Reference Bureau, www.prb.org/pdf06/06WorldDataSheet.pdf.

Though most developed countries in Europe and Japan have either fairly stable or declining populations, the US is still growing quite rapidly, mainly due to high levels of legal immigration, tolerance of illegal immigration, and the relatively high initial fertility levels of the new immigrants. US population at the current rate of expansion might attain about 420 million by 2050.[5] In all cases, developed countries depend on drawing resources from less developed ones. The financial, political, and cultural aspects of this are described as "under-development," which we might more accurately call *exploitation*. It is therefore not surprising that less developed countries retain a certain level of suspicion regarding anything the developed world might offer in the way of "aid and assistance." The problem is particularly acute for any concerted effort to control world population well in advance of fossil fuel depletion taking the decision out of human hands – through mass starvation. One aspect of the developed/underdeveloped nexus is that less developed countries often resent "solutions" offered, or imposed, by the richer countries.

Fortunately, the United Nations – with most countries of the world as members, including both haves and have-nots – has a fairly aggressive strategy to reduce world population growth rates. Information, education, and materials are offered to help people limit family sizes and live well with the children they do have. Declining infant mortality alone goes a long way towards lowering fertility. With fewer infant deaths, parents do not have to have so many babies in order for one or two to reach adulthood. This is one way the developed world can provide assistance without being accused of telling the less developed countries what to do – by funneling aid through the United Nations, representing all countries. (This will not work, of course, if the developed world dominates UN population policies, or withholds funds to support them.)

The UN Population Fund (UNFPA) and the UN Population Information Network have active programs to help countries control runaway population growth and to assess its causes. When releasing the UNFPA report titled *The State of World Population 2000*, its Executive Director, Dr. Nafis Sadik, said: "Millions of women are denied reproductive choices and access to health care, contributing each year to 60 million unwanted or mistimed pregnancies and some 500,000 preventable pregnancy-related deaths. Nearly half of all deliveries in developing countries take place without a skilled birth attendant present." According to the World Bank it was estimated in 2001 that 1.1 billion people had consumption levels below US$1 a day and that 2.7 billion lived on less than US$2 a day in 2001.[6] At least half of the current population of the world lives in poverty.

ENERGY AND POVERTY

Both natural resources and wealth are unequally distributed. Even the so-called "rich oil-exporter nations" are mainly low- or medium-income countries, using World Bank criteria (from US$400 to US$1,500 gross national product (GNP) per capita). Increasingly, as world economic rates decline, the growing dichotomy between the haves and have-nots leads to political instability, ethnic and community conflict, civil disturbance, and often armed conflict.

According to Energy Information Administration (EIA) projections[7] on current trends, world energy use could grow by up to around 73 percent by 2030, going from 446.7 quadrillion British thermal units (Btu)[8] in 2007 to 773.3 Btu. Population growth, industrial development, and increasing per capita energy use are the major determinants of energy demand growth. This returns us to the crises that will surely come as we pass through peak oil and enter a period of ever-declining supply. No immediate solution is likely, given the current global economic and political context.

One of the central "riddles" of development is how commercial energy supplies, which enable and maintain conventional economic growth, can be "taken out of the equation" without chaotic economic impacts. All conventional models of development are growth-based and growth-seeking – yet if all countries consumed energy at US or even European rates, the depletion and then final exhaustion of remaining oil and gas reserves would take only 20 or 30 years, instead of about 60 years. The environmental impacts due to a world population of around 9 billion persons consuming fossil energy resources at US or even European per capita rates of today would most certainly be catastrophic. The challenge therefore is to *reduce consumption* in the rich nations, and for poor countries and people to escape poverty without crippling their economies, the biosphere, climate, or other natural support systems.

The greatest challenge facing us is to improve the lot of the poor *without* greatly increasing the inefficient and polluting use of fossil fuels. Reduced consumption by the rich countries and energy conservation are two immediate options. Development of new and renewable energy sources is another. But these must be accompanied by the halting, and then reversal, of world population growth.

HOW MANY PEOPLE?

Joel Cohen wrote a well researched and scholarly book titled *How Many People Can the Earth Support?*[9] He tracks the historical answers to his primary question from the earliest one listed, in 1679, to the year 1994. Estimates range from the very small (500 million people, by Ehrlich) to the ultimate extreme of a world inhabited by human beings at a density of 500 persons per square *meter* of land surface, in buildings with an outer skin temperature of 2,000°C, this "heat limit" world population being about 1 thousand million billion. Most variations in estimates made over time are due to varying levels of scientific knowledge (lower for earlier forecasters), and to widely varying assumptions regarding what physical or biological factors set final limits on human population.

In a 2004 book[10] I responded to the question of how many people the earth should support. My answer was that this will depend on the kind of world you want. Will all people live at similar standards of living, or will this vary widely, as is currently the case? What is the most important failure factor in human population growth? Is it loss of food, energy, or capacities for waste removal? Or does the limit come from the spreading of killer diseases? More than 40 years ago, we added the possibility of global thermonuclear war annihilating all major (and even minor) centers of population.

David Pimentel of Cornell University wrote a short article estimating the maximum carrying capacity of the planet, assuming that all people would attain 1999 average US standards of consumption.[11] His estimate called for a substantial *reduction* in the current human population, to about 1 or 2 billion, arguing that the earth would be incapable of supporting the current world population at American levels of affluence. This conclusion is similar to Wackernagel's – in other words, that supporting the world's current human population at US levels of affluence would require not one but *three* earths.

Virginia Abernethy commented on these estimates, as follows:

> It is small wonder that numerous students of carrying capacity, working independently, conclude that the sustainable world population, one that uses much less energy per capita than is common in today's industrialized countries, is in the neighborhood of 2 to 3 billion persons. [We should also] note the congruence with Watt's projection of rapidly declining population size near the end of the Oil Interval. The absence of cheap, versatile, and easily used sources of energy, and other resources, seems likely to change the quality of human life and may even change, for many, the odds of survival.[12]

We are currently supporting a much larger population than Pimentel's estimate, because most of the current 6 billion people have a substantially lower standard of living than the one he used for making his projection. In addition, our *temporary* sources of energy (depleting fossil fuels) are supporting populations which, without energy-extended resource domains, would not be able to continue existing. Any estimate of long-term maximum supportable human numbers must assume the absence of substantial quantities of fossil-fuel resources (see Chapter 14, this volume). A substantial increase in human population above the current 6-billion-plus mark might be possible, but only: (1) at the expense of other life-forms with which humans compete, but on which humans also depend; (2) by lowering the overall material standard of living (and ecological impact) of the current human population; or (3) by finding ways to reduce human impact on other life-forms while human population and affluence continue to grow. The last of these alternatives can only be considered wishful thinking, but is grist to the mill of the "limit denial industry," the technological optimists, and growth-seeking economists believing in market-triggered "human inventiveness."

All estimates of large increases in the world's population are constrained (while generally denying this) in ignoring human rights, rejecting biodiversity, depreciating the aesthetic and cultural aspects of natural environments and non-human life-forms, and reducing human beings to mindless goods-consuming units. If we maintain the industrialized world while allowing the rest of the world to grow substantially in numbers, the consequence is to doom much of that "other world" to perpetual misery, while clinging on to the industrialized way of life as desperately and as long as possible. This scenario would very probably lead to a major population crash, as energy and other limits were quickly reached and then exceeded.

It is clear that we are facing serious moral and ethical issues as we approach the end of the "petroleum interval." To focus on the moral questions, I postulate three different scenarios:

- **Business as Usual Scenario.** The industrialized nations continue as they are, increasing energy efficiency and switching gradually to renewable energy sources, even as their populations grow slightly, primarily from immigration, and their per capita energy consumption levels remain high. The underdeveloped nations continue with current trends of intensive agricultural development, urbanization, and industrialization, using more energy and resources, and generating more pollution in the process. This continues until the combined stress on resource availability and the environment yields a catastrophic collapse of social and political systems. Worldwide economic collapse will most certainly be an encroaching condition in the process. The present conflict over resources, especially oil, escalates to the level of world wars. Through these wars and their impact on food production, and assuming they are non-nuclear, world population may be reduced drastically. However, "collateral damage" to the environment and resource-supplying systems will be large, precipitating a die-off. Several billion people will die in a short period, before sustainable levels are reached.

- **Selfish Nation Scenario**. The industrialized nations, seeking to maintain their affluent ways of living and materialistic perspectives, isolate themselves as much as possible from the rest of the world. Sources of oil in weak, remote nations are appropriated by economic and (increasingly) military force. The less developed world is shut out and cut off from the benefits of affluent living. Powerful industrial nations try by all (necessarily military) means to perpetuate this extreme dichotomy. Populations of the less developed nations decline sharply, with the breakdown of political, social, and cultural structures.
- **Humanitarian Scenario**. A massive program to reduce population growth is initiated, and strongly supported, with financial and economic resources from the developed countries. Developed nations embark on a global program of radical resource efficiency and increasing utilization of new and renewable energy sources (NRES). All perfected NRES are made available, with financial and economic support, to less developed nations. Energy-wasteful transport and urban–industrial practices are rapidly and radically modified, to reduce energy consumption and environmental impact. World population stabilizes in a few decades, thereafter declining gradually to approximately 4 billion by 2050.

Which of these scenarios we will produce is a matter of vast importance. If we choose the last of them, considerable public education, debate, and enlightened decision making will be required. In choosing amongst the possible future scenarios, we are facing serious moral and ethical decisions. The issues must be debated extensively, globally, and rapidly. Since humanity has taken over control of Spaceship Earth, we simply must decide where we want the great ship headed.

THE MORAL DILEMMA

Do the industrial countries owe anything to those in underdeveloped countries living lives of misery? Will the industrial world be willing to alter its own system to benefit the starving billions elsewhere? How much should the industrialized countries be willing to sacrifice for the sake of the underdeveloped world? Is it moral to conclude that we should not make such sacrifices, or is the very question born of a fallacious understanding of what it takes to live well? These are serious questions demanding thoughtful answers.

It is difficult to motivate people to change to a lifestyle they see as less desirable than the one they currently enjoy. I believe that ways can be found to live better with much less energy, and with lower material consumption, than currently enjoyed in the developed world. We can find non-polluting, low-energy ways to live, eat, sleep, dream, and enjoy life. By slowing our pace and simplifying our lives, we can learn to live well and happily without destroying the earth which is our home. Positive visions for humanity's future are needed to provide the essential motivation for change. Under any scenario, population growth must stop – ideally by human instigation, to prevent nature from having to do it for us. It is in fact an open question whether even the UN population projections are feasible and possible *without* bringing the world to the brink of near extinction, because of excessive damage to many critical components of the natural life-support systems of the biosphere. Finding out if this is possible is an experiment we are now pursuing. My children, and those of any parent, will see how it comes out.

NOTES

1. M. Wackernagel and W. Rees, *Our Ecological Footprint – Reducing Human Impact on the Earth*, New Society Publishers, Gabriola Island, BC, Canada, and Philadelphia, PA, USA, 1996.
2. M. Wackernagel, N.B. Schulz, D. Deumling, A. Callejas-Linares, M. Jenkins, V. Kapos, C. Monfreda, J. Loh, N. Myers, R. Norgaard, and J. Randers, "Tracking the Ecological Overshoot of the Human Economy," *Proceedings of the National Academy of Sciences*, USA, vol. 99, no. 4 (July 9, 2002), pp. 9266–71.

3. C. Cheslog, "New Report Outlines the Ecological Footprint of 146 Nations," Redefining Progress Media Release, 2002, www.rprogress.org/media/releases/021125_efnations.html, accessed December 31, 2002.
4. W. Youngquist, *GeoDestinies*, National Book Company, Oregon, 1997.
5. US Census Bureau's middle-range projection, cited at Floridians for a Sustainable Population, www.flsuspop.org/.
6. World Bank, "Understanding Poverty," http://tinyurl.com/9eeyk, accessed December 26, 2007.
7. EIA, "International Energy Outlook 2007," http://www.eia.doe.gov/oiaf/ieo/excel/ieohtab_1.xls.
8. One Btu is equal to the amount of heat energy required to raise the temperature of one pound of liquid water by 1 degree Fahrenheit. One Btu is equal to approximately 251.9 calories or 1,055 joules.
9. J.E. Cohen, *How Many People Can the Earth Support?*, W.W. Norton, New York and London, 1995.
10. R. McCluney, *Humanity's Environmental Future*, SunPine Press, Cape Canaveral, Florida, 2004.
11. D.L. Pimental, "How Many Americans Can the Earth Support?" *Population Press*, vol. 5, no. 3 (March/April 1999).
12. V.D. Abernethy, "Population and Environment Assumptions, Interpretation, and Other Reasons for Confusion," *Where Next? Reflections on the Human Future*, Board of Trustees, Royal Botanical Garden, London, 2000.

14
Renewable Energy Limits

Ross McCluney

The more we get out of the world the less we leave, and in the long run we shall have to pay our debts at a time that may be very inconvenient for our own survival.

Norbert Wiener

LIMITS

It would be impossible to supply enough renewable energy to support our current population and its level of activities, let alone those projected to the middle of the twenty-first century. It might be possible, however, at some indefinite time in the future, to completely sustain a smaller population, or a series of smaller populations, with renewables, as was done for centuries prior to the fossil fuel revolution that began in Britain in the eighteenth century.

Between the current situation and a hypothetical future with a small, sustainable population lies a necessary world of conjecture. I treated this topic in the preceding chapter, "Population, Energy, and Economic Growth: The Moral Dilemma."

TO THE RENEWABLES

Many people hope we can avoid the threatened difficulties by switching from petro-energy to solar energy, backed by increased energy conservation. The possibilities seem promising. New technologies include wind and solar powered electric generating stations, solar heating systems, ocean energy systems of several kinds, and possibly geothermal energy.

Though energy conservation and solar energy are to many our great hopes for the future, these technologies do have limitations.

FUTURE ENERGY TRANSITIONS

There have been several energy transitions in the past; from wood to coal, then to petroleum, natural gas, and nuclear. Depletion of forests has been a recurring factor in the decline of civilizations.[1] Transitions tended to be forced on societies through scarcity of old fuels, such as wood and charcoal, or, in the case of coal, because the pollution problems were so appalling and because high-quality sources close to the surface were becoming scarce.[2] Petroleum was very versatile, easy to transport, and initially plentiful and more accessible than coal. It permitted the rise of the automobile and air travel, making it the most popular fuel ever. Gas was a by-product of oilfields, usually discarded until the mid-1970s when petroleum costs rose. Nuclear energy has been adopted in countries which historically had few other sources of fuel and wished to avoid dependency on imported oil, particularly after the oil shock of 1973. France and India are notable examples. Nuclear energy is, however, not terribly versatile. It is really only safely available as electricity and there are growing doubts about uranium supply and the capacity of technology to exploit other fissionable sources, such as thorium.

Hydropower, where it involves large dams, has the disadvantage of requiring massive infrastructure and enormous territory to be inundated. Geothermal, although clean, is relatively rare on the surface and commercial exploitation in the earth's depths remains technically problematic.[3] Neither hydropower nor geothermal can be transported, except as the end-product electricity, so their availability for new energy generation is geographically limited.

We are finding that the next energy transition, away from the fossil fuels, will be toward a variety of different sources, none of them as abundant, versatile, reliable, or transportable as coal or oil. We seem to be looking largely at the relocalization of energy sources. Because of our huge fossil fuel populations, which have burgeoned in giant cities, which will almost certainly not be possible without fossil fuels, the new transition will be more difficult than the previous ones because it will involve more people than ever before, most of them entirely dislocated from primary production and therefore alienated from the source of food and other important materials.

Solar energy

"Passive solar energy" refers to opportunities to take advantage of available sunlight without artificially collecting or storing it. The siting and design of buildings, including the materials used, by making the most of available sunlight, can greatly reduce the need to use other fuels for heating and cooling.

Direct use of energy from solar or sky radiation

Solar or sky radiation[4] has been proposed as the great new replacement for fossil fuels, but it is substantially less concentrated than petroleum. To store it requires conversion to some other energy form, such as heat, electricity, or chemical energy. This is difficult and costly, requiring the processing of metals and glass to manufacture complex hardware which lasts only decades and presents pollution and disposal problems. In some cases it even has serious environmental drawbacks.

Solar thermal power plants are big, but – relative to other types of power plants – they're space efficient. Large-scale solar thermal technology known as concentrating solar power plants (CSP) seem to use a lot of land,[5] but when looking at electricity output versus total size, they use less land than hydroelectric dams (including the size of the lake behind the dam) or coal plants (including the amount of land required for mining and excavation of the coal). While all power plants require land and have an environmental impact, the best locations for solar power plants are on land, such as deserts, for which there might be few other uses (except for the occasional desert species that might be living there).

In a partially "solarized" society, one can envisage using our remaining, but declining, non-renewable energy sources for storage – to fill in the times when solar is not available, at least temporarily avoiding the need for solar storage. In a more fully solarized economy, electricity generated with solar-derived stored energy could provide the backup. The serious problems of biogas and methanol from (solar powered) crops are dealt with extensively by Alice Friedemann in this volume.[6]

It is true that a copious quantity of energy arrives from the sun each day. It falls all over the earth, but to harvest it directly, in sizeable quantities, means diverting it from uses in nature. Massive use of solar energy will require alteration of vast areas of the land and water surfaces of the planet, changing biosphere systems in the process.

In addition to being dilute, solar conversion systems currently require fossil fuels to manufacture them. "A major solar energy cost component is the cost of non-renewable resources of oil, natural gas, coal, and nuclear energy consumed in producing and constructing the systems for solar heating and solar electric plants."[7]

Proper assessment of a proposed solar technology should include a determination of the system's net energy production, that is, the magnitude of the solar-energy-derived output minus the non-renewable energy drawn from the earth, needed to make and operate the solar energy system.

Calculations show that the net energy of solar collection and distribution systems in some cases is negative. Also, at the end of their useful lives these systems must be dismantled and recycled (with additional expenditures of energy).[8]

If the net energy output of a solar technology is negative, logic leads to the question, "Why bother?" In such a case, wouldn't it be better to use the fossil energy directly rather than lock it up in an inadequately producing solar energy system?

This is a controversial topic. Even if the calculations are correct for some situations, there can be value in storing present day (less expensive) fossil energy in solar collection devices, as a hedge against future depletion. Continuing with this argument, since we are going to use up fossil fuels anyway, why not invest that energy in the manufacture of renewable energy systems, so they can go on producing power when the fossil fuels are depleted? Ultimately we would like to remove the non-renewable energy inputs from the manufacture of solar energy systems altogether.

Solar "breeding" system

This leads to the idea of a solar "breeding" system – using solar energy in the mining and processing of ore and the manufacture of solar energy systems, thereby reducing or eliminating the fossil fuel subsidy. Solar energy systems produced by such a system will be strong net energy gainers. For such a strategy to be successful, the solar powered mining and manufacturing industry must be completed before the fossil fuel sources are gone (or before they become exorbitantly expensive). This might enable the establishment of a society based solely on solar energy, using solar energy alone to recycle worn-out solar systems. Ultimately we might even be able to develop a completely sustainable process not requiring the extraction of further minerals or fossil fuels from the earth to keep it going, as long as requirements for "fresh" inputs are continually reduced.

Solar pollution

The issue is not just about the non-renewable energy subsidy required to make and operate solar energy systems. The degree of environmental destruction associated with an energy consuming or producing system of any kind is also critical. As Baron pointed out in 1981, "Even more serious would be the impact upon public health and occupational safety if solar energy generates its own pollution when mining large quantities of energy resources and mineral ores."[9] Some solar energy manufacturing processes produce toxic or otherwise undesirable waste products which have to be recycled, discarded, or otherwise rendered benign. Clearly, we'll have to pick and choose amongst the solar alternatives to find the least environmentally impacting ones, and work hard to improve all the rest.

Solar limits

There are physical limits to the production of energy from direct solar radiation. In the absurd limit, we clearly could not cover all available land area with solar collectors. A more reasonable limit would be to fill existing and future rooftops with solar collectors. From data provided by the US Energy Information Administration, I estimated the total combined commercial and residential building roof area in the United States in the year 2000 at 18 billion square meters. From a National Renewable Energy Laboratory website, I found that the approximate annual average quantity of solar energy falling on a square meter of land area in the United States is about 4.5 kilowatt hours (kWh) of energy per square meter of area per day. Multiplying this by 365 days in a year and by the 18 billion-square-meter roof area figure, yields the total energy received by rooftops in this scenario: 2.46 × 1,013 kWh per year, or 84 Quads[10] per year. This is somewhat below the 102 Quads per year US primary energy consumption figure. Not all roof area is usable, however. Roofs sloped away from the sun's strongest radiation, shaded by trees and other buildings, having interfering equipment, or being insufficiently strong to support solar equipment, are either not practical or not possible for this utilization.

The conversion from primary to end-use energy is not perfectly efficient in either the renewable or the non-renewable cases. Both the 102 and the 84 Quads of primary energy must therefore be reduced when converting them to actual end-use energy. It is difficult to determine accurate average conversion efficiencies for all technologies in both categories, but they are not likely to be widely different. Thus the conclusion should remain valid that meeting total US energy needs with 100 percent direct solar energy would require about every single square foot of roof area of all commercial and residential buildings in the country.

Since most of the current roofs were neither designed nor built to carry the loads of (and wind loading on) solar collectors filling them, nor are they all exposed adequately to the sun, it is very unlikely that we could achieve the goal of a 100 percent solar economy in this manner. For every hectare of existing rooftops which cannot be filled with solar collectors, an equivalent hectare would have to be found elsewhere. Renewable energy from other sources would also be needed. If US population continues to grow, pressure will continue mounting to expand developed land areas into what are currently agricultural and wilderness areas. If the plan is to convert as much as possible of the US energy economy to direct solar energy, solar collector farms will join in the competition for new lands to be opened up for this development.

In order not to have to convert agricultural or natural habitat areas to areas for engineered solar production, one would have to find other, already developed areas for erecting these solar collectors, such as street and highway corridors and parking lots. While the collector-shaded areas might be attractive to persons having to drive and park in the hot sun, it is probably not economically feasible under current financing conditions. A number of other objections to this possibility can be expected, leading to continued pressure to convert agricultural and wilderness areas to solar production "farms."

What about the deserts?

A common reaction to the problem of finding areas for large solar energy collection systems in forests, on farms, or in developed areas is to point out the vast "unused" desert areas around the globe, suggesting that these would be good places for solar collectors. Surely some desert areas can be used for renewable energy technology, but there are limits. Deserts are not devoid of wildlife; they contain varieties of flora and fauna, adapted over millions of years to desert conditions. There is a limit to how much desert we can cover with solar collectors. In spite of this, I suspect that this limit is very large, that we have a long way to go before we near saturation on environmental grounds. Of course this presumes some level of acceptance of vast solar energy farms in the deserts along with the extinction of some species in the process. It remains to be seen what that level of acceptance is.

Hydrogen gas in a solar system

Hydrogen gas is a much-touted means of energy storage. It can be solar-produced, through solar powered electrolysis of water. It is clean-burning and non-toxic. However, it is the lightest of the chemical elements and difficult and costly to store and concentrate. Research is in process to find ways of storing and releasing hydrogen chemically, avoiding the need for expensive, heavy, and potentially dangerous high-pressure storage tanks. If the problems can be overcome, our hopes for hydrogen as a portable fuel may be realized, but this will be neither easy nor inexpensive.

INDIRECT SOLAR – RENEWABLE ENERGY TECHNOLOGIES, POSSIBILITIES, AND PROBLEMS

In addition to direct collection of solar energy several indirect sources of this important resource exist. They include the wind, powered by differential heating of the earth's surface, ocean currents (produced by a similar mechanism), hydroelectric, powered by solar-powered water evaporation and condensation into rivers, ocean thermal energy conversion (based on solar-heated surface layers of the tropical oceans), and ocean waves, driven by the wind and carrying energy with them as they approach the shoreline.

Waves and thermal-driven currents

Waves and thermal-driven currents offer a degree of natural solar concentration. Tidal currents are also concentrated in some locations. Solar-derived wind, pushing the sea over large distances, increases wave heights and their energy content. This energy can be extracted downwind, where the waves are most intense. Thermal currents can be focused between land masses, thereby concentrating the speed and energy content of the moving fluid. Let's take a look at each of these renewable technologies.

Wind power

Wind power has now become economically viable for areas experiencing adequate average wind speeds. Due to the difficulty of finding onshore sites and other factors, wind turbines are also being sited offshore.

It is expected that within the next few years, wind parks with a total capacity of thousands of megawatts will be installed in European seas – the equivalent of several large, traditional coal-fired or nuclear power stations. Plans are advancing for wind parks in Swedish, Danish, German, Dutch, Belgian, British, and Irish waters. Outside Europe there is also serious interest.[11]

Many wind proponents claim that wind farms on land can cohabit with agriculture and that leasing such land can be a valuable source of income for the farmer. A substantial number of wind turbines have been installed in windy desert areas, and more are expected. Wind is generally variable in its speed, however, so it is not the most suitable source for what is called "baseload" electricity generation, that nonvarying core power component that forms the backbone of electric utility operations.

Though the problems are relatively minor at present, wind turbines have been shown to be hazardous to birds in some locations. If the landscape is covered with these large devices, more problems can be expected. If the areas covered by them increase a hundred-fold, however, opposition on visual and amenity grounds can be expected to increase. Noise pollution from some turbines is likely to remain as an impact for local human settlements.

The hazards due to offshore wind "farms" include dangers for navigation and disturbance of local marine fauna. Often when human structures are placed near the coastline, they are heralded as "artificial reefs" capable of increasing populations of a variety of marine species, generally considered a good thing, but a possible problem when huge areas are concerned.

Ocean currents

Ocean currents, such as the Florida Current, the part of the Gulf Stream flowing northward past the Florida peninsula, carry enormous quantities of kinetic energy in their motion. There have been several proposals to develop this resource, to place ocean turbines in the strongest of currents and feed the energy generated to population centers onshore. According to Practical Ocean Energy Management Systems, Inc., "The first large ocean-system proposal is for a 2.4-mile system that would link Samar and Dalupiri islands in the Philippines. The Dalupiri project is now estimated to cost $2.8 billion, produce 2,200 megawatts (MW) at tidal peak and offset 6.5 million tons of carbon dioxide a year."[12]

Calculated roughly, in the Florida Current, the kinetic energy transported through cross-sectional area by a fluid of known mass density is the product of its kinetic energy per unit mass of moving fluid and the mass flow rate through that cross-sectional area. The energy flow rate, per unit area, is proportional to the cube of the current speed. All the kinetic energy contained in the flowing water cannot be usefully extracted or the flow would cease. It should be possible to extract enough energy to slow the stream by about 50 percent or so. In this case, approximately 88 percent of the available kinetic energy would be extracted.

The Florida Current, between Miami and the Bahama Islands, shows potential for electrical generation. The Gulf Stream flows northward through the straits at a speed ranging from two knots at the edges to over four knots in the middle of a 20-nautical-mile (37 km) width off Miami, yielding an approximate average kinetic energy per unit cross-sectional area transported by this current in the order of 2,000 watts per square meter (W/m^2). If we assume an 88 percent conversion efficiency (slowing the current by a factor of two in velocity) and that we extract this energy from the surface down to a depth of 10 meters over the 37 km width of the current (370,000 m^2), then the total electrical power output for a 100 percent efficient electricity generator would be in the order of $0.88 \times 2000 \times 370000 = 651$ MW. If we choose a 30-meter depth, the power generation would be three times larger, or 1.9 gigawatts (GW), a large electrical generation capacity.[13]

Ocean currents are one of the largest untapped renewable energy resources on the planet. Preliminary surveys show a global potential of over 450 GW.[14]

Undersea currents

Undersea current energy extractors have much potential in regions where conditions are right, but I believe it doubtful that the economic value of the electricity that might be generated, though large, would be sufficient to offset the huge costs of construction, including anchoring underwater structures in strong current in deep water, and dealing with whatever environmental consequences might be produced.

Tidal energy

Tidal energy can be extracted by placing turbines or other current energy extractors in or across the mouths of estuaries experiencing large tidal excursions. Energy in the flow of ocean water in and out of the estuary can be extracted and turned into electricity. A working power plant of this type is located in France. It produces 240 MW of power via a "barrage" across the estuary of the River Rance, near Saint Malo in Brittany. The plant went online in 1966 and supplies about

90 percent of Brittany's electricity. This is a fairly unique installation. It is doubtful that it could be duplicated at reasonable cost in many places around the world.

For tidal differences to be harnessed into electricity, the difference between high and low tides should be at least five meters, or more than 16 feet. There are only about 40 sites on the planet with tidal ranges of this magnitude.

Tides of less magnitude, however, could be used to produce usable power. Turbines placed under the water, grounded on the bottom, could allow shipping to pass overhead while still generating power. Currently, there are no operational tidal turbine farms of this type. European Union officials have, however, identified 106 sites in Europe as suitable locations for such farms. The Philippines, Indonesia, China, and Japan also have underwater turbine farm sites that might be developed in the future. The costs of such massive undersea structures are likely to be high. It is a real question whether future increases in petroleum costs will justify extensive exploitation of the tidal resource.

Tidal energy, although having a large potential, is restricted to the estuarine areas experiencing significant tidal swings. Tidal power plants that dam estuaries can impede sea-life migration, and silt buildups behind such facilities can impact local ecosystems adversely. Tidal "fences" may also disturb sea-life migration. Newly developed tidal turbines may prove ultimately to be the least environmentally damaging of the tidal power technologies because they do not block migratory paths; however the future economic feasibility of these huge underwater structures, anchored to the bottom, has not been proven.

Ocean thermal energy

Ocean thermal energy is another potential source. The sun heats the surface waters of the tropical oceans, making them considerably warmer than water at great depths. It is possible to run a heat engine between these two thermal regions. A working fluid, such as ammonia, placed in a partial vacuum, is evaporated by heat taken from the warm surface water; the evaporated gas expands against a large turbine, making it spin to produce electricity. The gas is condensed after passing through the turbine by cooling it with deep ocean water. Since the temperature difference between the two heat reservoirs is modest, in comparison with a fossil fuel steam power plant, the efficiency of conversion to electricity is quite low. On the other hand, the "fuel" (solar-heated water) is free for the taking, so that such a plant should eventually pay for itself over time. So far no commercial ocean thermal energy conversion plant has been built, mainly for reasons of too long payback times. As energy prices increase, payback times shorten, generally leading to the opening of new markets.

Ocean waves

Ocean waves carry a substantial amount of energy. According to the US Department of Energy, the total power of waves breaking on the world's coastlines is estimated at 2–3 billion kW (2–3 terawatts (TW)). In favorable locations, wave energy density can average 65 MW per mile of coastline.[15] Of course, due to environmental problems, land use conflicts, hazards to navigation, and other reasons, only a small fraction of this power can be extracted for human use. Wave power devices extract energy directly from surface waves or from pressure fluctuations below the surface.

Wave power cannot be harnessed everywhere. Wave-power-rich areas of the world include the western coasts of Scotland, northern Canada, southern Africa, northern Australia, and the northeastern and northwestern coasts of the United States. Wave energy utilization devices have been built and operated in a number of locations around the world. Several European countries have programs to deploy wave energy devices.

Ocean wave energy has a very large energy potential, but also many environmental and technological hurdles to overcome. Impacts include potentially drastic hydrological effects of structures on shoreline and shallows ecology; potential navigation hazards; noise pollution with possible disruption of marine mammal communications; visual pollution especially near land; severe land-use conflict impacting recreational and other uses. Installation and possibly operation of ocean wave energy conversion devices and the laying of electrical cables will damage and affect species on the sea bed and in the water column. There is also the possibility of offshore sites significantly altering coastal wave regimes with unpredictable effects.

Hydroelectric

Hydroelectric power generation is used extensively around the world. United States hydropower facilities can generate enough power to supply 28 million households with electricity, the equivalent of nearly 500 million barrels of oil per year. The total US hydropower capacity – including pumped storage facilities – is about 95 GW. There are probably a number of sites around the world where rivers can be dammed and hydropower developed, but the environmental impacts can be huge. Thus, the potential for new energy from this source is limited.

In addition to the flooding of valleys and destruction of upland habitat, hydropower technology may cause fish injury and mortality from passage through turbines, as well as detrimental effects on the quality of downstream water. Severe land-use conflicts arise from social and ecological impacts of flooding large areas.

Geothermal

Geothermal energy is abundant and includes radioactive heat which is common in certain deep rocks. Although some geothermal is easy to access, much is not and requires expensive technology. This subject is discussed elsewhere in this volume.[16]

ENVIRONMENTAL IMPACT OF RENEWABLES

At the current minuscule level of renewable energy generation, what little environmental consequences might result from renewable systems is pretty much a drop in the bucket compared with those of fossil fuels. However, as fossil fuels switch roles with renewables, the relatively minor impacts experienced now can grow to a substantial size.

The UK Department of Trade and Industry described some of the problems with renewables:

> Despite their benefits, renewables present important issues that need to be addressed. The main constraints on their use are the costs of the energy they produce and the local environmental impacts of renewable energy schemes. Currently, the cost of energy from renewables is generally higher than that produced by "conventional" energy sources. However, as renewables become more established and the benefits of mass production take effect, the gap will reduce. Indeed, in the case of wind power and some other technologies, this is already happening.[17]

As energy prices rise with the decline of oil, the cost effectiveness of the renewable options should increase. The impacts of renewable energy technologies, I believe, can only increase.

How much might all potential sources supply?

If we combine the energy production potentials for all the sources mentioned above, we could supply all the world's energy needs, probably several times over. There are limits, however, as discussed below.

CONSERVATION LIMITATIONS

Energy conservation is another important strategy. It goes hand in hand with renewable energy systems, as a means of keeping capital investment costs low. Improved energy efficiency in transportation systems is also possible, but not without massive redesign of transport vehicles and systems.

Under any hypothesis, energy conservation must be a component of any strategy for meeting future energy needs.

Reducing one's energy needs also reduces the cost and environmental impacts of a solar system, making it a more socially and environmentally viable option.

The problem is that efficiency, by itself, is just a multiplier. If the number of people and their per capita demand for energy is allowed to continue rising, these factors will swamp whatever efficiency gains one obtains.

ECONOMIC AND SOCIAL CONDITIONS

In the EU, fossil and nuclear power are publicly subsidized to the tune of €15 billion per year. In addition, the European taxpayer picks up the environmental and human health bill for acid rain, for nitrogen oxide emissions, for particulates, and for the "natural" disasters caused by climate change, itself triggered by fossil fuel burning.

In the US, government subsidies to the fossil fuel industry are not matched for renewables.

Economic analysis and financial costing remains tilted against renewable energy – the playing field is far from level. Existing electricity systems and laws often make it extremely difficult for renewables to gain fair access to national markets. Perhaps the most serious problem is that current "free" market practices offer no mechanisms for including non-monetary benefits, along with the energy savings of a given technology. Such technologies are therefore undervalued in the market and underutilized.

CONCLUSION

Growth in world population and rising expectations for plentiful energy will put great pressure on developers to capture as much of each resource as possible, perhaps with terrible environmental consequences. Judicious use, however, of all the profiled renewable energy sources might meet a smaller world population's energy needs with minimal environmental impact, thereby making the goal of a fully sustainable society realizable.

NOTES

1. D.R. Montgomery, *Dirt: The Erosion of Civilizations*, University of California Press, Berkeley and Los Angeles, 2007.
2. B. Freese, *Coal: A Human History*, William Heinemann Ltd, London, 2005.
3. See Sheila Newman's chapter "Geothermal," this volume.
4. Solar radiation reaches the earth's surface either by being transmitted directly through the atmosphere ("direct solar radiation"), or by being scattered or reflected to the surface ("diffuse sky radiation"). National Snow and Ice Data Center (NSIDC), USA, http://nsidc.org/arcticmet/factors/radiation.html.
5. For an idea of what is entailed, see "Solar Energies Technologies Program," US Department of Energy, www1.eere.energy.gov/solar/csp.html.
6. See Alice Friedemann's chapter "Peak Soil," this volume. And: "We are facing an epic competition between the 800 million motorists who want to protect their mobility and the two billion poorest people in the world who simply want to survive. In effect, supermarkets and service stations are now competing for the same resources" (Lester Brown, "Ethanol Could Leave the World Hungry," *Fortune* Magazine, August 16, 2006).
7. S. Baron, *Mechanical Engineering*, vol. 103 (1981), p. 35.
8. Peter Knudson, *New Mexico Monthly Bulletin*, vol. 3 (1978), p. 16.
9. Baron, *Mechanical Engineering*, p. 35.
10. One Quad is equal to 1,000,000,000,000,000 British thermal units (Btus).
11. "Concerted Action on Offshore Wind Energy in Europe CA-OWEE," *Renewable Energy World*, vol. 5 (2002), p. 28.
12. Practical Ocean Energy Management Systems (POEMS), Inc., "Ocean Current Technical FAQ," 2002, www.poemsinc.org/currentFAQ.html.
13. See Harris B. Stewart, Jr., "Proceedings of the MacArthur Workshop on the Feasibility of Extracting Usable Energy from the Florida Current," Palm Beach Shores, Florida, February 27 to March 1, 1974.
14. POEMS, "Ocean Current Technical FAQ."
15. Office of Energy Efficiency and Renewable Energy, US Department of Energy, "Ocean Topics," www.eren.doe.gov/RE/ocean.html, last update December 18, 2002.

16. Sheila Newman's chapter "Geothermal," this volume, deals more extensively with geothermal energy resources.
17. UK Department of Trade and Industry (DTI), "Sustainable Energy Programmes – Introduction," 2002, www.dti.gov.uk/renewable, accessed December 29, 2002.

15
Peak Soil

Alice Friedemann

Biofuel is marketed by powerful and popular concerns as the answer to pricey gasoline, global warming, and fears about overdependence on imported fuel in the US today. Since conservation and reduction of consumption don't sell in a growth-oriented consumer economy, potential yields of biofuel from the edible grains of food crops have been hugely exaggerated and further bulked up with the forecast that "cellulosic" biofuels are just around the corner.

For instance, United States Democratic presidential candidate John Edwards promises to

> force oil companies to install ethanol pumps at a quarter of their service stations and require automakers to build cars that can run on biofuels ... [and cause US consumption of ethanol to] soar to 65 billion gallons a year by 2025 [with a $13 billion fund to] make cellulosic ethanol cheaper than corn-based ethanol.[1]

The US Department of Energy (DOE) proclaims on its website[2] that

> Biomass use strengthens rural economies, decreases America's dependence on imported oil, avoids use of MTBE or other highly toxic fuel additives, reduces air and water pollution, and reduces greenhouse gas emissions ... [and] tomorrow, biorefineries will use advanced technology such as hydrolysis of cellulosic biomass to sugars and lignin and thermochemical conversion of biomass to synthesis gas for fermentation and catalysis of these platform chemicals to produce slates of biopolymers and fuels.

Grain processor Archer Daniels Midland (ADM) spent three decades lobbying for ethanol to be used in gasoline. Today, ADM makes record profits from ethanol sales and government subsidies.[3] POET[4] and various associated industries are also doing very well.[5] And today, the US DOE wants to replace 30 percent of petroleum consumption by having biomass supply 5 percent of the nation's electricity, 20 percent of transportation fuels, and 25 percent of chemicals by 2030.[6]

US government subsidies to large agribusiness companies give the public the impression that biofuels are coming along. In reality, only large grain merchants like ADM and POET, and select industries upstream and downstream,[7] including the oil industry which receives 51 cents[8] per gallon of subsidy money to blend ethanol with gasoline, and of course the automobile industry and all its dependents which need the *corn*ucopia myth to justify continued production, benefit from the rainforest destruction, topsoil and aquifer depletion involved in growing crops on an industrial scale for food and fuel.

As well as competing directly with food production, growing plants to make fuels such as ethanol, biodiesel, and butanol is tremendously ecologically destructive and unsustainable.

Currently all industrial/commercial biofuels are made from the edible grains of food crops, but biofuel enthusiasts talk about using plant "waste," such as cornstalks, wood, grasses, and other inedible plant parts, to make "cellulosic biofuel." Biologically speaking, the concept of plant "waste" is grotesque. These hard, indigestible parts of plants are actually vital to the structure of soil. They are made of lignocellulose and plants evolved them over hundreds of millions of years to grow tall and prevent animals from eating them.

Large agribusiness corporations may consider the inedible parts of plants "waste," but crop residues, grasses, and trees recycling into the soil are essential to maintaining the top six inches of soil to grow the food crops on which our lives depend.

As well as the fact that soils cannot really spare lignocellulose to keep North Americans in the style to which they have become accustomed, the conversion of lignocellulose to fuel takes about five times more energy than it produces.[9]

Biofuels to replace petroleum is an agribusiness get-rich-quick scheme that will bankrupt our topsoil and turn the land into a desert. This is elementary soil biology, but because the voice of soil scientists is not being heard, most people do not realize that if you destroy the soil, you can't grow biomass. If you cannot grow biomass, not only can you not make ethanol to fuel cars, you cannot grow food either.[10]

Fuels from biomass are not sustainable and there isn't enough biomass in America to make significant amounts of energy because essential inputs like water, land, fossil fuels, and phosphate ores are limited.

SOIL SCIENCE 101 – THERE IS NO "WASTE" BIOMASS

Long before there was "peak oil," there was "peak soil." Productivity drops off sharply when topsoil reaches six inches or less, the average crop root zone depth.[11] Crop productivity continually declines as topsoil is lost and residues are removed.[12]

On average it takes about 100 years to form an inch of topsoil. Erosion from poor farming and other abuses can remove centuries of topsoil in less than a decade.[13] Industrial agriculture and the expansion of suburbs to crop land is eroding topsoil far faster than it is being formed now.

Erosion is happening ten to twenty times faster than the rate topsoil can be formed by natural processes.[14] The natural, geological erosion rate is about 400 pounds of soil per acre per year[15] but economic redefinitions of erosion have obscured signs which should arouse strong concern. For instance, the US Department of Agriculture (USDA) has defined erosion as the average soil loss that could occur without causing a decline in long term productivity. In fact, on *over half* of America's *best* crop land, the erosion rate is 27 times the natural rate, 11,000 pounds per acre.[16] Widely accepted values of how to assess erosion do not even account for the time it takes for the deep layers – the subsoil – to develop from rock.[17]

Erosion removes the most fertile parts of the soil.[18]

When you feed the soil with organic matter, you're not feeding plants; you're feeding the biota – that is, the living organisms in the soil – for soil is *alive*.

Underground creatures and fungi break down fallen leaves and twigs into microscopic bits that plants can eat, and create tunnels through which air and water can infiltrate. In nature there are no elves feeding (fertilizing) the wild lands. When plants die, they're recycled into basic elements and become a part of new plants. It's a closed cycle. There is no biological waste. "Biowaste" is a construct of theoretical economics which has no real-life existence.

Soil creatures and fungi are the immune system for plants. They protect them against diseases, weeds, and insects – when this living community is harmed by agricultural chemicals and fertilizers, even more chemicals are needed in an increasingly vicious and harmful cycle which eventually kills the living soil through starvation, toxicity, and chemical transformation.[19]

There's so much life in the soil; there can be ten "biomass horses" underground for every horse grazing on an acre of pasture.[20] Just a tiny pinch of earth could have 10,000 different species[21] – millions of creatures, most of them unknown. If you dived into the soil and swam around, you'd be surrounded by thousands of miles of thin strands of mycorrhizal fungi that help plant roots absorb more nutrients and water.[22] As you swam along, plant roots would tower above you like trees as you wove through underground skyscrapers.

Plants and creatures underground need to drink, eat, and breathe just like we do. An ideal soil is half rock, and a quarter each water and air. When tractors plant and harvest, they crush the life out of the soil, as underground apartments collapse 9/11-style. The tracks left by tractors in the soil are the erosion route for half of the soil that washes or blows away.[23]

Many plants want animals to eat their seed and fruit to disperse them. Some seeds only germinate after going through an animal gut and coming out in ready-made fertilizer. Seeds and fruits are easy to digest compared to the rest of the plant; that's why all of the commercial ethanol and biodiesel are made from the tasty parts of plants: the grain, rather than the stalks, leaves, and roots.

But plants don't want to be entirely devoured. They've spent hundreds of millions of years perfecting structures that can't easily be eaten. Be thankful plants figured this out, or everything would be mown down to bedrock.

The total removal of all parts of plants, including the cellulosic parts we don't eat, (husks, stalks, and roots), misleadingly termed "residues," deprives the soil of water, carbon, and other nutrients[24] which artificial fertilizers do not supply. It is like expecting a human body to live off vitamin pills whilst depriving it of food. These "residues" also build new soil and create a water-retaining protective cover that prevents soil erosion, allowing air and water to reach plant roots and permitting underground aquifers to recharge.

Removing residues leads to lower crop production and ultimately deserts. Growing plants for fuel will accelerate the already unacceptable levels of topsoil erosion, soil carbon and nutrient depletion, soil compaction, water retention, water depletion, water pollution, air pollution, eutrophication, destruction of fisheries, siltation of dams and waterways, salination, loss of biodiversity, and damage to human health.[25]

Corn is the most widely grown and subsidized row crop in America, and the favorite for biofuel production, attracting government subsidies. Row crops cause 50 times more soil erosion than sod crops,[26] or more.[27] Corn uses more water, agrichemicals, and fertilizer than most crops.[28] High corn prices are leading to continuous corn-cropping, rather than the healthy method of rotating crops which replace nitrogen in the soil and assist erosion control.[29] Continuous cropping, particularly of corn, increases eutrophication by 189 percent, global warming by 71 percent, and acidification by 6 percent.[30] Eutrophication removes the oxygen in water, destroying fisheries. Acidification lowers the pH of the soil, ultimately preventing most crops from being grown.

Farmers increasingly pressure to farm ecological buffer-zones and conservation areas so as to take advantage of rocketing corn prices.[31]

Natural gas and fertilizers in agriculture

Fertilizers draw on 28 percent of the energy used in agriculture.[32] They depend on natural gas as both feedstock and fuel for the artificial manufacture of nitrogenous fertilizers. So let's get this straight. Fertilizers are made from and with natural gas which we're dumping on crops to grow them for biofuel. We're going to take the biomass waste away, which means we'll have to add even more fertilizer. How, exactly, does that lessen our dependence on fossil fuels?

Carbon gases

Fertilizers do not provide the ecosystem services that organic matter does. Organic matter slows erosion and fixes carbon in the soil. Soils contain 3.3 times the amount of carbon found in the atmosphere, and 4.5 times more carbon than is stored in all the earth's vegetation.[33] If we're going to slow global warming down, we need to return residues to the soil, not try to make fuels out of them.

Land clearing for biofuel cultivation

Energy farming, wherever it is practiced, is playing a huge role in deforestation, reducing biodiversity, water and water quality, and increasing soil erosion. Fires to clear land for palm oil plantations are destroying one of the last great remaining rainforests in Borneo, spewing out so much carbon that Indonesia is third behind the United States and China in releasing greenhouse gases. Orangutans, rhinos, tigers, and thousands of other species may be driven to extinction.[34] Borneo palm oil plantation lands have grown 2,500 percent since 1984.[35] Soybeans grown as a row crop cause even more erosion than corn and suffer from all the same sustainability issues. The Amazon is being destroyed by farmers growing soybeans for food[36] and fuel.[37] This rainforest destruction makes an important contribution to global warming.

Energy crops

As well as advocating the harvesting of cellulosic material from food crops, biofuel proponents propose "Energy Crops," crops grown specifically for their biofuel value. Some are traditional food crops, like corn and sugar cane. Non-food energy crops include tall perennial grasses such as switchgrass and miscanthus, which cause less erosion and need less fertilizer than food energy crops. For every plant-hope dashed by practical and scientific criticism, biofuel proponents will pull another out of the hat, including trees. But energy crops suffer from the same problems all plants have: robust lignocellulose which is more useful to soil than to our fuel needs.

Silence from soil scientists

The problems of erosion, over-reliance on fertilizers, and soil depletion are well known. Yet soil scientists seem to be absent from the biofuel debate. A poll of 35 soil scientists indicated to me that they fear that expressing their views will cost them their jobs.[38] Here is what one commented anonymously:

> Government policy since WWII has been to encourage overproduction to keep food prices down (people with full bellies don't revolt or object too much). It's hard to make a living farming commodities when the selling price is always at or below the break even point. Farmers have had to get bigger and bigger to make ends meet since the margins keep getting thinner and thinner. We have sacrificed our family farms in the name of cheap food. When farmers stand to make a few bucks (as with biofuels) agricultural scientists tend to look the other way.

Understanding of the importance and function of soil goes back centuries and this debate is decades old.[39] In 1911, the USDA wrote that

> With the passing years, the soil became more compact, droughts were more injurious, and the soil baked harder and was more difficult to handle. Continuous corn culture has no place in progressive farming ... it is a short-sighted policy and is suicidal on lands that have been long under cultivation.[40]

Unrealistic targets

The DOE Biomass "1 billion tons of waste biomass" plans to harvest 400 million tons of corn and wheat residues fail to take into account the most basic facts about plants.[41] For instance, only about 0.5 percent of a plant can be harvested sustainably every year. Only 51 million tons of corn and wheat residues could be harvested without causing erosion, not 400 million tons.[42] Fifty-one million tons of residue could make about 3.8 billion gallons of ethanol per annum – less than 1 percent of United States energy needs. And this relatively modest potential is constantly declining. The United States lost 52 million acres of cropland between 1982 and 2002.[43] At that rate, all of the cropland will be gone in 140 years.

The government believes that the US can produce, from one plant source or another, such as trees, switchgrass, and city waste, a billion tons of biomass "waste" to make cellulosic biofuels and chemicals, and to generate electricity.[44]

In truth, there isn't enough biomass to replace 30 percent of our petroleum use. The potential biomass energy is miniscule compared to the fossil fuel energy we consume every year, about 105 exajoules (EJ; 1 EJ is equal to 10^{18} joules) in the US. If you burned every living plant and its roots, you'd have 94 EJ of energy and we could all pretend we lived on Mars. Most of this 94 EJ of biomass is already being used for food and feed crops, and in wood for paper and homes. Sparse vegetation and the 30 EJ in root systems are economically unavailable – leaving only a small amount of biomass unspoken for.[45]

ENERGY RETURNED ON ENERGY INVESTED

Studies that show a small positive energy gain for ethanol of 1.2 would show a negative return if the "by-product" were not counted.[46] The transformation of a bushel of corn plus two gallons of water and some yeast into ethanol produces 18 pounds of ethanol, 18 pounds of carbon dioxide (CO_2) and 18 pounds of "by-product." By-product furnishes 20 percent of cattle-feed and 5–10 percent of pig-feed, at most.[47] Plans for corn-ethanol to provide 10 percent of US energy would result in an amazing 37 times oversupply of by-product stock-feed requirements for all US livestock.[48] This low-quality stock-feed supplement succumbs to mold and fungi after four to ten days and its bulkiness and volume makes widescale distribution energy costly, so that spreading it back in its traditional place on the soil would constitute an energy cost absurdity.

Harvesting, storage, and transportation

All plants, including energy crops, have low density compared to fossil fuels. If you try to compact them, that takes energy, and they're still low density. Hay bales are like mattresses – you can only get so many on a truck, and you can't force them into a pipeline, which would be far less expensive. If you try to compact them further by turning them into pellets, it takes so much energy that you are entering negative energy land.

Plants aren't concentrated – they grow diffusely and require a great deal of energy to harvest and deliver to the refinery – a biorefinery needs plants delivered from the surrounding 7,000 or more square miles for a 2,000 ton/day refinery. Plants are hard to store. They rot and turn into mulch or can catch on fire. Storing them wet adds weight, leading to higher transportation costs and high water use. All plants succumb to pests and disease, especially if grown as monocrops.

Peer-reviewed comparative studies in major scientific publications find that biofuels cost more fuel to produce and transport than is gained from the plant material after processing.[49]

In summary, plants are hard to make into fuels

There is not enough water for people, industry, and biofuel refineries now, but by 2100, the US Census projects potentially 1.1 billion people in the United States.[50]

Time explains why renewable energy provides such low-energy yields compared to non-renewable fossil fuels. The more work left to nature, the higher the energy yield, but the longer the time required. Although coal and oil took millions of years to form into dense, concentrated solar power, all we had to do was extract and transport them.[51]

With every step required to transform a fuel into energy, there is less and less energy yield. For example, to make ethanol from corn grain, which is how all ethanol is made now, corn is first grown to develop hybrid seeds, which next season are planted, harvested, delivered, stored, and pre-processed to remove dirt. Dry-mill ethanol is milled, liquefied, heated, saccharified, fermented, evaporated, centrifuged, distilled, scrubbed, dried, stored, and transported to customers.[52]

Biorefineries need to be enormous for economies of scale – 100 acres of hay stacked 25 feet high.

You can't do this at home – biofuels need to be pure or combustion engine life may be shortened, and used within three months before microbes chew them up. Some of the gasohol made in the 1980s had so much water in it that gasohol got a bad name, which is why it's called ethanol in it's new reincarnation.

As a systems architect and engineer, the author looks at projects from start to end, trying to identify the failure points. The US Department of Energy Biomass Roadmaps and the Energy Biosciences Institute Proposal have taken a similar approach and identified the barriers to cellulosic fuels. Business proposals for biofuels have lacked this systemic and scientifically testable approach and government policy is responding more to business than to science. The result could be a disaster for the United States with impacts on all those countries which model their policy on US policy and those which depend on US food exports, such as Japan. (See Anthony Boys's Chapter 22 in this volume.)

In business you're limited by money; in science, you're limited by the laws of physics and thermodynamics. When it comes to biofuels, you're also limited by *ecosystems*. To grow plants sustainably, the soil ecosystem and water supply need to be taken into account. No wonder many of the issues with cellulosic ethanol aren't discussed – there's no way to express the problems in a soundbite.

The success of cellulosic ethanol depends on finding or engineering organisms that can tolerate extremely high concentrations of ethanol. Augenstein argues that such creatures would already exist if they were possible. Organisms have had a billion years of optimization through evolution to develop a tolerance to high ethanol levels.[53] Someone making beer, wine, or moonshine would have already discovered this creature if it could exist.

It may not be possible to reduce the complex cellulose digesting strategies of bacteria and fungi into micro-organisms or enzymes that can convert cellulose into biofuels in giant steel vats, especially given the huge physical and chemical variations in feedstock. The field of metagenomics is trying to create a chimera from snips of genetic material of cellulose-digesting bacteria and fungi. That would be the ultimate Swiss Army-knife microbe, able to convert cellulose to sugar and then sugar to ethanol.

There's also research to replicate termite gut cellulose breakdown. Termites depend on fascinating creatures called protists[54] in their guts to digest wood. The protists in turn outsource the work to multiple kinds of bacteria living inside of them. This is done with adenosine triphosphate energy[55] and architecture (membranes) in a system that evolved over millions of years. If the termite could fire the protists and work directly with the bacteria, that probably would have happened 50 million years ago. This process involves many kinds of bacteria, waste products, and other complexities that may not be reducible to an enzyme or a bacteria.

Jay Keasling, Director of Physical Biosciences at Lawrence Berkeley National Lab (LBNL), proposes to do the above in a synthetic biology factory. You'd order up the biological bits you need to create a microbial machine the way electronics parts are obtained at electronics stores. At UC Berkeley on April 21, 2007, he said this could also be used to get past the 15 percent concentration of ethanol that pickles microorganisms, which results in a tremendous amount of energy being used to get the remaining 85 percent of water out. Or you could use this technology to create a creature that could convert miscanthus and switchgrass to create biofuels that can be put in pipelines and burned in diesel engines.[56]

Biologists roll their eyes when reductionist physicists pat them on the head and tell little ol' biology not to worry, living organisms can be reduced to atoms and enzymes, just take a piece of algae here, a bit of fungi or bacteria there and *voila*! – a new creature that produces vast volumes of biofuels quickly. But biology is a messy wonderment. Creatures exist within food webs and don't reproduce well if surrounded by their own toxic wastes. The research may be well intentioned, but public policy shouldn't assume synthetic biology is a "slam dunk."

Although it is possible these microbes can be engineered, as they are in biotechnology labs where drugs are produced, the challenge is different when it comes to making energy producing microbes. The microbes need to convert lignocellulose into a biofuel quickly. In nature, the process is very slow. And the energy to keep them alive, remove toxic waste products, and so on, must not exceed the energy ultimately produced, which isn't a consideration when making pharmaceutical products.

But meanwhile we're stuck with corn and ethanol, which in the end must be delivered to the customer. Since ethanol can't be delivered cheaply through pipelines, but must be transported by truck, rail, or barge,[57] this is very expensive for the coastal regions. Alaska and Hawaii have managed to get out of having to add ethanol to gasoline, but California's Senator Feinstein has not been able to do the same.

The whole cellulosic ethanol enterprise falls apart if the energy returned is less than the energy invested or even *one* of the major stumbling blocks can't be overcome. If there isn't enough biomass, if the residues can't be stored without exploding or composting, if the oil to transport low-density residues to biorefineries or deliver the final product is too great, if no cheap enzymes or microbes are found to break down lignocellulose in wildly varying feedstocks, if the energy to clean up toxic by-products is too expensive, or if organisms capable of tolerating high ethanol concentrations aren't found – if these and other barriers[58] can't be overcome, then cellulosic fuels are not going to happen.

If the obstacles can be overcome, but we lose topsoil, deplete aquifers, and poison the land, air, and water, what a Pyrrhic victory.

Scientists have been trying to turn plants into fuel and other products for over 30 years. The research is intrinsically worth doing, because perhaps someday plants can help replace the fossil fuels used in half a million products.

However, we need to wind back our lifestyles tremendously, as soon as possible, because biofuels cannot begin to replace the volumes stored in our depleting fossil reserves.

Corporations exist to make profits for shareholders. Destruction of fisheries, topsoil, aquifers, and other "externalities" are not considered or even acknowledged. Ethanol production is profitable *only* because of government subsidies. Ethanol production would end if the subsidies to ethanol and gasoline refineries were stopped. Without subsidies biofuels would not be profitable. But the subsidies aren't likely to stop soon because there are far too many Congressmen from farm states who depend on this money recycling back into their campaign funds.

Meanwhile, we're betting civilization on getting biofuels to work, because we need a liquid fuel that can be burned in the millions of existing combustion engines we rely on for agriculture and transportation. It would take decades to replace our infrastructure with electrical engines, decades we don't have with oil depletion looming. But there should also be a Plan B to make a U-turn back to an agrarian society. We shouldn't bet the farm on a biofueled future.

NOTES

1. "Edwards Launches Plan to Boost Ethanol Use," Yahoo News, May 31, 2007, http://news.yahoo.com/s/nm/20070531/pl_nm/usa_edwards_energy_dc_2, accessed June 28, 2007.
2. A site promoting the DOE's Biomass "Multi Year Program Plan," www1.eere.energy.gov/biomass/, accessed 1 June, 2007.

3. A. Barrionuevo, "A Bet on Ethanol, With a Convert at the Helm," *New York Times*, October 8, 2006.
4. "Broin is POET," www.poetenergy.com/news/media.asp, accessed July 13, 2007.
5. http://tinyurl.com/ypdaxc, accessed July 13, 2007.
6. "Multi Year Program Plan 2007–2012," August 31, 2005, http://tinyurl.com/2jgq9l, and "Roadmap for Agriculture Biomass Feedstock Supply in the United States," November 2003, http://devafdc.nrel.gov/pdfs/8245.pdf, both from the Office of Energy Efficiency and Renewable Energy, Biomass Program, US Department of Energy, accessed July 13, 2007.
7. Such as ADM and POET, DuPont, Broin, Deere and Company: see, for instance, www.poetenergy.com/news/showRelease.asp?id=25&year=&categoryid=.
8. Michael O'Hare, "Subsidies are the Wrong Road to Biofuels," *San Francisco Chronicle*, July 30, 2006; Phillip Brasher, "Study: Ethanol Subsidies Total at Least $5 Billion," *DesMoines Register*, October 25, 2006.
9. J. Benemann and D. Augenstein, "Whither Cellulosic Ethanol?" *The Oil Drum*, August 16, 2006, www.theoildrum.com/story/2006/8/15/13634/6716, accessed July 13, 2007.
10. Loss of topsoil has been a major factor in the fall of civilizations. You end up with a country like Iraq, formerly Mesopotamia, where 75 percent of the farm land is a salty desert. B. Sundquist, *A Global Perspective, the Earth's Carrying Capacity*, "Topsoil Loss – Causes, Effects, and Implications," http://home.alltel.net/bsundquist1/se0.html, accessed May 6, 2005; W. Lowdermilk, *Conquest of the Land Through Seven Thousand Years*, USDA-SCS, 1948; J. Perlin, *A Forest Journey: The Role of Wood in the Development of Civilization*, Harvard University Press, Cambridge, Mass., 1991; C. Ponting, *A Green History of the World: The Environment and the Collapse of Great Civilizations*, Penguin Books, London, 1991.
11. Sundquist, "Topsoil Loss – Causes, Effects, and Implications: A Global Perspective."
12. David Montgomery, *Dirt: The Erosion of Civilizations*, University of California Press, Berkeley and Los Angeles, 2007, pp. 155–6, 161–77.
13. Mahdi Al-Kaisi, "Soil Erosion, Crop Productivity and Cultural Practices," Iowa State University, May 2001; B.C. Ball et al., "The Role of Crop Rotations in Determining Soil Structure and Crop Growth Conditions," *Canadian Journal of Soil Science*, vol. 85, no. 5 (2005), pp. 557–77; H. Blanco-Canqui et al., "Rapid Changes in Soil Carbon and Structural Properties Due to Stover Removal from No-Till Corn Plots," *Soil Science*, vol. 171, no. 6 (2006), pp. 468–82; Board on Agriculture (BOA), *Soil Conservation: An Assessment of the National Resources Inventory*, Volume 2, National Academies Press, Canadian Society of Soil Science, Manitoba, 1986; V. Calviño et al., "Corn Maize Yield as Affected by Water Availability, Soil Depth, and Crop Management," *Agronomy Journal*, vol. 95 (2003), pp. 275–81; A.J. Franzleubbers et al., "Agricultural Exhaust: A Reason to Invest in Soil," *Journal of Soil and Water Conservation*, vol. 61, no. 3 (2006), pp. 98–101; A.S. Grandy et al., "Do Productivity and Environmental Trade-Offs Justify Periodically Cultivating No-Till Cropping Systems?" *Agronomy Journal*, vol. 98 (2006), pp. 1377–83; J. Johnson et al., "Characterization

of Soil Amended with the By-Product of Corn Stover Fermentation," *Soil Science Society of America Journal*, vol. 68 (2004), pp. 139–47; J. Johnson, D. Reicosky, R. Allmaras, D. Archer, and W.W. Wilhelm, "A Matter of Balance: Conservation and Renewable Energy," *Journal of Soil and Water Conservation*, Iowa, vol. 61 (2006), pp. 120–5A; J. Miranowski, "Impacts of Productivity Loss on Crop Production and Management in a Dynamic Economic Model," *American Journal of Agricultural Economics*, vol. 66, no. 1 (1984), pp. 61–71; J. Power et al., "Residual Effects of Crop Residues on Grain Production and Selected Soil Properties," *Soil Science Society of America Journal*, vol. 62 (1998), pp. 1393–7; V. Sadras, "Quantification of Grain Yield Response to Soil Depth in Soybean, Maize, Sunflower, and Wheat," *Agronomy Journal*, vol. 93 (2001), pp. 577–83; F. Troeh and L.M. Thompson, *Soils and Soil Fertility*, 6th edition, Blackwell, Ames, Iowa, 2005; W. Wilhelm, M.F. Johnson, J.L. Hatfield, W.B. Voorhees, and D.R. Linden, "Crop and Soil Productivity Response to Corn Residue Removal: A Literature Review," *Agronomy Journal*, vol. 96, no. 1 (January–February 2004), pp. 1–17.

14. D. Pimentel, "Soil Erosion: A Food and Environmental Threat," *Environment, Development and Sustainability* (Journal) (February 2006); D. Pimentel et al. (Editorial) "Green Plants, Fossil Fuels, and Now Biofuels," *American Institute of Biological Sciences* (November 2006).
15. Troeh and Thompson, *Soils and Soil Fertility*. Some is due to farmers not being paid enough to conserve their land, but most is due to investors who farm for profit. Erosion control cuts into profits.
16. "Soil Erosion," *Conservation Resource Brief No. 0602*, Natural Resources Conservation Service, US Department of Agriculture, February 2006.
17. Troeh and Thompson, in *Soils and Soil Fertility*, p. 400, argue that the tolerable soil loss (T) value is set too high, because it is based only on the upper layers – how long it takes subsoil to be converted into topsoil. T ought to be based on deeper layers – the time for subsoil to develop from parent material or parent material from rock. If they are right, erosion is even worse than NRCS figures.
18. "About the Wind Erosion Research Unit," www.weru.ksu.edu/new_weru/about/about.shtml, US Agricultural Research Service, accessed July 13, 2007.
19. D. Wolfe, *Tales from the Underground: A Natural History of Subterranean Life*, Perseus Publishing, Cambridge, Mass., 2001.
20. T. Hemenway, *Gaia's Garden*, Chelsea Green, Vermont, 2000.
21. D. Wardle, "Ecological Linkages between Aboveground and Belowground Biota," *Science*, vol. 304 (2004), pp. 1629–33. The June 2004 issue of *Science* calls soils "The Final Frontier."
22. E. Pennisi, "The Secret Life of Fungi," *Science*, vol. 304 (2004), pp. 1620–23.
23. Wilhelm et al., "Crop and Soil Productivity Response to Corn Residue Removal: A Literature Review."
24. R.G. Nelson, "Resource Assessment and Removal Analysis for Corn Stover and Wheat Straw in the Eastern and Midwestern United States – Rainfall and Wind-Induced Soil Erosion Methodology," *Biomass and Bioenergy*, vol. 22 (2002), pp. 349–63, 2002; A. McAloon et al., "Determining

the Cost of Producing Ethanol from Cornstarch and Lignocellulosic Feedstocks," www.ethanol-gec.org/information/briefing/16.pdf, NREL, October 2000; J. Sheehan et al., "Energy and Environmental Aspects of Using Corn Stover for Fuel Ethanol," *Journal of Industrial Ecology*, vol. 7, nos 3–4 (2003), pp. 117–46.

25. E. Tegtmeier and M.D. Duffy, "External Costs of Agricultural Production in the United States," *International Journal of Agricultural Sustainability*, vol. 2, no. 1 (2004).
26. P. Sullivan, "Sustainable Soil Management: Soil Systems Guide," ATTRA, http://attra.ncat.org/attra-pub/PDF/soilmgmt.pdf, May 2004.
27. Mahdi Al-Kaisi, "Soil Erosion: An Agricultural Production Challenge," *Integrated Crop Management*, IC-484(19) (July 24, 2000), www.ipm.iastate.edu/ipm/icm/2000/7-24-2000/erosion.html. The soil between rows can wash or blow away. If corn is planted with last year's corn stalks left on the ground (no-till), erosion is less of a problem, but only about 20 percent of corn is grown no-till. Soy is usually grown no-till, but has insignificant residues to harvest for fuel.
28. M. Padgitt et al., "Production Practices for Major Crops in U.S. Agriculture, 1990–97," www.ers.usda.gov/publications/sb969/sb969.pdf, Resource Economics Division, Economic Research Service, USDA, September 2000; D. Pimentel, "Ethanol Fuels: Energy Balance, Economics and Environmental Impacts are Negative," *Natural Resources Research*, vol. 12 (2003), pp. 127–34.
29. Known respectively as "nitrogen fixing" and "sod crops." Nitrogen is a major component of artificial fertilizers but it is also readily and naturally available from certain crops.
30. S.E. Powers, "Quantifying Cradle-to-Farm Gate Life-Cycle Impacts Associated with Fertilizer Used for Corn, Soybean, and Stover Production," www1.eere.energy.gov/biomass/pdfs/37500.pdf, NREL, May 2005.
31. B. Tomson, "For Ethanol, U.S. May Boost Corn Acreage," *Wall Street Journal*, February 8, 2007.
32. M.C. Heller and G.A. Keoleian, "Life-Cycle Based Sustainability Indicators for Assessment of the U.S. Food System," www1.eere.energy.gov/biomass/pdfs/37500.pdf, University of Michigan, 2000.
33. R. Lal, "Soil Carbon Sequestration Impacts on Global Climate Change and Food Security," *Science*, vol. 304 (June 11, 2004), pp. 1623–6; R. Lal, "Soil Erosion Impact on Agronomic Productivity and Environmental Quality," *Critical Reviews in Plant Sciences*, vol. 17 (1998), pp. 319–464.
34. George Monbiot, "The Most Destructive Crop on Earth is No Solution to the Energy Crisis," *Guardian*, December 6, 2005, www.guardian.co.uk/science/story/0,3605,1659469,00.html.
35. P. Barta and J. Spencer, "As Alternative Energy Heats Up, Environmental Concerns Grow," *New York Times*, October 8, 2006; P. Barta, "Crop of Renewable 'Biofuels' Could Have Drawbacks," *Wall Street Journal*, December 5, 2006.
36. "Farming the Amazon," *National Geographic*, January 2007, p. 40ff.
37. J. Olmstead, "What About the Land? A Look at the Impacts of Biofuels Production, in the U.S. and the World," *Grist Environmental*

News and Commentary, December 5, 2006, http://grist.org/news/maindish/2006/12/05/olmstead/.

38. www.energyskeptic.com/Peak_Soil.htm.
39. R. Sampson, *Farmland or Wasteland: A Time to Choose. Overcoming the Threat to America's Farm and Food Future*, Rodale Press, Pasadena, 1981; W. Jackson et al., "Impacts on the Land in the New Age of Limits," *Land Report* No. 9:20, 1980; C. Smith, "Rotations in the Corn Belt," *Yearbook of the Department of Agriculture*, 1911.
40. Smith, "Rotations in the Corn Belt."
41. One ton equals 2000 lb, or 907.15 kg. "DOE Billion Ton Vision, Biomass as Feedstock for a Bioenergy and Bioproducts Industry: The Technical Feasibility of a Billion-Ton Annual Supply," http://feedstockreview.ornl.gov/pdf/billion_ton_vision.pdf, USDA DOE Biomass Plan, August 31, 2005, and "Multi Year Program Plan 2007–2012," US Department of Energy, Office of the Biomass Program, Energy Efficiency and Renewable Energy, US DOE, August 31, 2005.
42. Nelson, "Resource Assessment and Removal Analysis for Corn Stover and Wheat Straw in the Eastern and Midwestern United States – Rainfall and Wind-Induced Soil Erosion Methodology."
43. National Resources Inventory 2002 Annual NRI, www.nrcs.usda.gov/technical/land/nri02/landuse.pdf, April 2004.
44. "DOE Billion Ton Vision," and "Multi Year Program Plan 2007–2012."
45. T. Patzek, "The Real Biofuels Cycles," *Science*, vol. 312 (June 26, 2006), p. 1747, supporting material online; T. Patzek, "A Statistical Analysis of the Theoretical Yield of Ethanol from Corn Starch," *Natural Resources Research*, vol. 15, no. 4 (2006); D. Pimentel and T.W. Patzek, "Ethanol Production Using Corn, Switchgrass and Wood; Biodiesel Production Using Soybean and Sunflower," *Natural Resources Research*, vol. 14, no. 1 (2005).
46. A.E. Farrell, R.J. Plevin, B.T. Turner, A.D. Jones, M. O'Hare, and D.M. Kammen, "Ethanol Can Contribute to Energy and Environmental Goals," *Science*, vol. 311 (January 27, 2006), pp. 506–8, http://rael.berkeley.edu/ebamm/FarrellEthanolScience012706.pdf, accessed March 6, 2008.
47. A. Trenkle, "Integration of Animal Agriculture with the Bioeconomy," www.bioeconomyconference.org/images/Trenkle,%20Allen.pdf, Department of Animal Science, Iowa State University, 2006; J. Shurson et al., "Value and Use of 'New Generation' Distiller's Dried Grains with Solubles in Swine Diets," Department of Animal Science, University of Minnesota, Alltech 19th International Feed Industry Symposium Proceedings, 2003.
48. M. Giampietro et al., "Feasibility of Large-Scale Biofuel Production," *BioScience*, vol. 47, no. 9 (1997), pp. 587–600.
49. Patzek, "The Real Biofuels Cycles"; Patzek, "A Statistical Analysis of the Theoretical Yield of Ethanol from Corn Starch"; Pimentel et al., "Ethanol Production Using Corn, Switchgrass and Wood; Biodiesel Production Using Soybean and Sunflower."
50. Benemann and Augenstein, "Whither Cellulosic Ethanol?"
51. Howard T. Odum, *Environmental Accounting. EMERGY and Environmental Decision Making*, John Wiley & Sons, New York, 1996.

52. McAloon et al., "Determining the Cost of Producing Ethanol from Cornstarch and Lignocellulosic Feedstocks."
53. Benemann and Augenstein, "Whither Cellulosic Ethanol?"
54. Richard L. Howey, "You Want to Bring THOSE into the House???" *Micscape Magazine*, March 2003, UK, www.microscopy-uk.org.uk/mag/artmar03/rhtermite.html.
55. Adenosine triphosphate (ATP) is the universal energy source of *all* creatures.
56. Singer, E., "A Better Biofuel," *MIT Technology Review*, April 3, 2007, http://www.technologyreview.com/Biotech/18476/.
57. B. Yacobucci, "Fuel Ethanol: Background and Public Policy Issues," www.khoslaventures.com/presentations/Fuel_Ethanol.pdf, CRS Report for Congress, 2006.
58. Such as those mentioned in the appendix to the author's longer article, "Peak Soil: Why Cellulosic and Other Biofuels are Not Sustainable and a Threat to America's National Security," www.energyskeptic.com/Peak_Soil.htm.

16
Notes on Terra Preta

Sheila Newman

Terra preta (or "agrichar") are kinds of super-productive soil. If they result from technology rather than accident, this technology might be summarized as an ancient one which maximizes carbon-loaded surfaces for soil-micro-organisms, once employed by South American Indians and others (for example, in north-eastern Thailand).[1] The carbon which so enriched these special soils was somehow provided by reducing wood to charcoal at relatively low temperatures, in smothered fires, in the presence of wastes from human settlements. Although it is not yet certain whether this was intentional or accidental, trying to reproduce this effect is the focus of much commercial, scientific, and individual research and discussion. An inch of terra preta has been estimated to have taken 25 years to accumulate, with buildups of several feet taking several thousand years.[2]

The basic theory at this stage is that the porous structure of the charcoal retains organic matter and water and is thus very hospitable to soil organisms. Some of these organisms include arbuscular mycorrhizal fungae which form fungal "mats," like a structural matrix throughout the forest floor. As well as assisting plants to acquire phosphorus and micronutrients from the soil, these mats themselves "fix" or store soil carbon. This function is obviously of interest to conservationists, corporations, and governments alike, as a mechanism for reducing greenhouse gases by sequestering carbon. In fact, terra preta samples have been shown to contain 70 percent more carbon than surrounding soils.[3] The process by which the fungal mats store carbon involves their producing large quantities of glomalin.[4] Glomalin is a "glycoprotein," storing carbon as a protein and as a carbohydrate. Glycoproteins occur in animals as well as plants. For instance, mucus is glycoprotein-rich. Containing 30–40 percent carbon, glomalin characterizes a kind of connective tissue in the soil community, chemically and physically binding organic to inorganic particles, forming soil granules. These sticky granules give form and shape to the soil, conserving the carbon and other nutrients in it. It takes high heat to destroy this connectivity, which may explain why terra preta is connected to low-heat charcoal.[5] This stickiness makes Terra preta exceptionably resistive to the leaching which affects other soils in high rainfall conditions.[6]

In fact, a mechanistic explanation of terra preta might invoke the role of micro-organisms as living "machines" in living soil-factories which form metabolic links between living and non-living matter, of which the output is forests of plants, trees, and animals, including humans. To speak of "machines" and "soil factories" is, however, reductionist and clunky, because it ignores the self-reproductive, cooperative, and other qualities of life itself. From Alice Friedemann's chapter, "Peak Soil," we know that soil is our most basic fuel. Soil processes are integral to whole ecosystems in which we are ourselves encapsulated and glomalin is implicated in recent explanations of the development of life on earth.[7] This statement will perhaps whet the reader's appetite to go and find out more.

It is a promising area of soil research, but, if we find out how to reproduce what the Amazonians benefited from, it is almost certainly only going to be the kind of soil technology which can be applied locally, in relocalization systems. Plans to develop this technology for profit to attract massive funding from global investors seem misguided for the same reasons that plans to industrialize biofuels by genetically designing super micro-organisms seem impractical. They are limited by the scale and complexity of ecosystems. It makes more sense to adapt our economies to ecosystems rather than to attempt to recreate and adapt ecosystems to our economies. In recent times the availability of plentiful fossil fuel has conditioned many to the idea that we can find the energy to reinvent anything. As well as being the motor for the creation of masses of industrial waste, this is probably an illusion which will vanish along with cheap fossil fuel supplies.

NOTES

1. David R. Montgomery, *Dirt*, University of California Press, Berkeley and Los Angeles, 2007, pp. 143–4.
2. Ibid.
3. E. Marris, "Putting the Carbon Back: Black is the New Green," *Nature*, vol. 442 (August 10, 2006), pp. 624–6, www.nature.com/nature/journal/v442/n7103/full/442624a.html; B. Glaser, "Prehistorically Modified Soils of Central Amazonia: A Model for Sustainable Agriculture in the Twenty-First Century," *Philosophical Transactions of the Royal Society B*, vol. 362 (2007), pp. 187–96.

4. http://en.wikipedia.org/wiki/Arbuscular_mycorrhiza.
5. "Wright knew she had something unique when it took up to 90 minutes in a heat-sterilizing autoclave to free glomalin from the hyphae ... With that level of persistence, we knew glomalin must stay in the soil, too." From an article about Sara E. Wright's discovery of the properties of glomalin in soils, "Glomalin – Soil's Superglue," www.ars.usda.gov/is/AR/archive/oct97/glomalin1097.htm.
6. Andrew B. Carey, "Aiming to Restore Forests, Evaluation with SER Criteria," *Northwestern Naturalist*, vol. 87, no. 1 (Spring 2006), www.fs.fed.us/pnw/pubs/journals/pnw_2006_biodiversity_005_carey.pdf.
7. Frank C. Landis, A. Gargas, and T.J. Givnish, "Relationships among Arbuscular Mycorrhizal Fungi, Vascular Plants and Environmental Conditions in Oak Savannas," *New Phytologist*, vol. 164 (2004), pp. 493–504; and http://en.wikipedia.org/wiki/Glomalin.

17
Nuclear Fission Power Options

Sheila Newman

NUCLEAR ENERGY

Burning fossil fuel gives us chemical energy, but what is atomic energy?

Einstein was the first to say that mass is energy. Mass is made up of atoms and represents an enormous form of energy if you can release it. In radioactivity atoms release their mass via fission or splitting in a natural process known as "decay." When a radioactive element splits it loses a very slight amount of its mass. That mass gets converted to heat energy. There are three major naturally occurring radioactive elements – uranium, thorium, and potassium.

Structure of atoms and radioactive behavior

Atoms are made up of non-reactive "neutrons" which, explained according to current dominant theory, are bound by a theoretical "weak nuclear force"[1] to "protons" and other neutrons. A cloud of electrons moves around these units. The neutrons have no electrical charge but the protons have a positive one, equal to a negative charge in each of the orbiting electrons. The electrons also have negligible mass compared with the nucleus, rather like the sun and the earth. In the case of the sun and the earth, the earth orbits the sun due to the sun's much greater mass, and gravity maintains this relationship. Down at the atomic level it is the electrostatic charge which maintains the atomic relationship. In this case, however, unlike the planets, electrons have equal mass to each other. The physical and chemical properties of the elements depend on the number of protons, that is, on the basic electrostatic "charge."

An element must have a certain combination of protons plus neutrons to be stable. As the nucleus gets larger the binding effect of the neutrons tends to be overcome by the strong repulsion between protons. So larger nuclei tend towards radioactive "decay."[2] In the "decay" process they eject single or combinations of neutrons, protons, electrons and photons in processes of upwards graduating radiation danger to humans, known as alpha, beta, and gamma. Highly dangerous at close range, gamma radiation consists of photons which are very short electromagnetic waves. They are capable of fatally altering chemical bonds at a molecular level in living cells.

On the way to a stable element, radioactive decay predictably produces slightly different forms of the elements, known as isotopes. These transformations cost mass and each successive isotope has a decreasing atomic weight, in a situation where the composition of a material is changing at sub-atomic level from one element to another. Uranium-238 begins with an atomic mass of 330. Gradually, over 9 billion years plus, it becomes the element lead, with 82 protons and an atomic mass of 207.2. The number of its protons defines an element and its "charge." The number of neutrons is the atomic number.

ENERGY USED FOR PRODUCTION OF NUCLEAR POWER (EROEI)

Separation (enriching) process

Uranium ore (yellowcake) contains approximately 0.1–0.25 percent of actual uranium oxides. Naturally occurring uranium tends to contain about 0.7 percent uranium-235 (U-235) and the rest uranium-238 (U-238). U-238 is not "fissile" but it is "fertile," which means it has the potential to become fissile U-235 in an artificial process of neutron bombardment (breeding). U-235 can maintain the fission process in a chain reaction without requiring supplementary neutron bombardment. Breeder reactors are designed to use extra neutrons to convert fertile fuels: U-238 and thorium (Th-232) into artificial fissile fuels: plutonium-239 (Pu-239) and uranium-233 (U-233).

The aim is to maximize the amount of fissile material available to the reactor. This often involves concentrating the fissile material in a process known as "separation."

The amount of energy that goes into the separation process is one of the most important factors to consider when evaluating energy output by a nuclear plant. Recent changes in method require far less energy.

CANDU reactors

A well-established reactor which does not use enriched fuel also exists. Originating in Canada, CANDU reactors[3] do not expend energy in separating fissile from non-fissile uranium isotopes. But their operation does require purchase of heavy water from an industry which separates the deuterium (D_2O), a naturally occurring isotope of hydrogen, from ordinary water. One CANDU uses about 360 tonnes of "heavy" water, which lasts the reactor for about 30 years with a loss of about 5 percent. D_2O occurs naturally in the ratio 1:4,500; thus D_2O is found at the level of about 1 in 20 million water molecules. 340,000 pounds of ordinary water is required to produce one pound of heavy water in the distillation process.[4] The price of heavy water is around US$300 per kg.[5]

Already during World War II the Germans used heavy water to moderate a reactor which used unenriched uranium. D_2O has an extra neutron and interacts with U-238 by slowing down the rate at which U-238 throws off neutrons, permitting a higher rate of collision-causing fission.[6] D_2O is not radioactive because, having only one neutron, it lacks the requisite instability. Radioactive tritium is, however, a by-product of processing uranium with D_2O.

CANDU proponents compare performance very favorably with all other models; pointing out that they do not have to shut down to refuel and that they can use different fissile materials, including thorium. Like all nuclear reactors, they produce and burn plutonium.

Now CANDU manufacturers are pioneering a new model which uses ordinary water and "slightly" enriched uranium, which raises questions regarding some of the above design sales points. Its reliability and safety record are used to promote it as a vehicle for recycling nuclear wastes.

Other conventional reactors

For non-CANDU reactors, the kind of enrichment process used makes a big difference to the up-front energy costs.

Understanding the difficulties of separating fissile from non-fissile could be helped by knowing that for the Manhattan Project at Oak Ridge, Tennessee, 60,000 workers worked for three years to separate 2 kg of U-235 from U-238 in two different buildings by gaseous diffusion in one and by ionization in another between 1943 and 1945.[7]

> For nuclear power, enrichment is clearly the key energy input where the older diffusion technology is used – it comprises more than half the lifetime total. However, with centrifuge technology it is far less significant than plant construction. There is an overall threefold difference in energy ratio between these two nuclear fuel cycle options.[8]

Enrichment accounts for almost half of the cost of nuclear fuel and about 5% of the total cost of the electricity generated. It can also account for the main greenhouse gas impact from the nuclear fuel cycle if the electricity used for enrichment is generated from coal. However, it still only amounts to 0.1% of the carbon dioxide from equivalent coal-fired electricity generation if modern gas centrifuge plants are used, or up to 3% in a worst case situation.[9]

Since uranium isotopes do not differ in their chemical behavior, enrichment techniques exploit their mass difference as a means for separating them.

The most prevalent methods are gaseous diffusion and gas centrifuge. Gaseous diffusion forces gaseous uranium hexafluoride (UF_6) under compression through semi-permeable membranes, managing to separate some molecules of U-235 from U-238. Gaseous diffusion needs 1,000 consecutive separation cascades to approximately 10 consecutive separation cascades required in gas centrifuge. Compression of UF_6 at the entry point of each cascade is energy-intensive, whereas gas centrifuge only needs electrical energy for the rotation of the cylinders plus heat to work. Gas centrifuge is taking over from gaseous diffusion for this reason, and now about 60–70 percent of reactors use centrifuge.[10] Depending on the composition of the original ore more or less work may be required to obtain the requisite mix approaching 3.5 percent U-235. In the current economic paradigm the cost of the electricity to drive the enrichment process would be weighed up against the cost of yellowcake. Amounts of enriched uranium are frequently described in terms of "Separative Work Units" (SWUs). Laser enrichment processes currently being worked on hold some promise of greater separation efficiency, but none are yet in operation.[11]

Breeder reactors

Most reactors round the world use uranium. A breeder reactor must be started with enriched uranium or some fissile substitute.

During the operation of a conventional nuclear power station over the years, via the reactions that occur in the atomic pile, the U-238 that is there gets converted into fissionable material which includes plutonium and other dangerously radioactive products. These products are increased in a breeder reactor. In conventional reactors moderators (or coolants) slow the neutron firing down so that the neutrons hit each other more easily and accelerate the natural rate of fission.

Without the moderator the reactor becomes a "fast breeder" with a bias towards the U-238 being converted and producing more fuel than it actually burns. In a fast breeder, the nuclear waste products which present such a problem in conventional reactors become more fuel. The aim is to make this a closed and remote controlled process.

The first fast breeder reactor (FBR) or fast neutron reactor was built in the US in 1951 with a tiny output of 0.2 megawatts electric (MWe) and operated until 1963, when it was succeeded by a 20 MWe one, a 66 MW one, a 20 MW one and "Fast Flux TF" which had a thermal output of 400 MW, from 1980 to 1993. The UK had a 15 MWe one from 1959 to 1977 and then a 270 MW one from 1974 to 1994. France built its first in 1966 with an output of 40 MW thermal (MWth), followed by "Phénix" in 1973 with 250 MWe (still in operation) and "Superphénix" in 1985 to 1998 with an output of 1,240 MWe. Germany had one very small one with an output of 21 MWe from 1977 to 1991; India has one with an output of 40 MWth, built in 1985. Japan's "Joyu" with 140 MWth was built in 1978; its Monju FBR at 280 MWe went from 1994 to 1996 and is currently closed. Kazakhstan's BN350 has been going since 1972 with an output of 135 MWe, half of which desalinates about 80,000 tonnes of water each year for the city of Aktau. Russia has had three FBRs: the first in 1959–71 reopened in 1973; the second, from 1979, produces 12 MWe; and the third, built in 1981, with an output of 600 MWe, is the largest still running (with assistance from a US supervisory crew), but it has had a lot of problems with liquid sodium coolant and other leaks, involving long periods out of action. France's Superphénix was the biggest in the world, but it was closed down due to safety problems associated with sodium leaks. Monju in Japan was also closed due to safety concerns.

Significant commercial success seems to have been elusive so far, but there are international ambitious plans for "Generation IV" FBRs of various designs, including thorium-based ones. The low cost of uranium is often offered as an explanation for the failure of thorium technology to find the necessary finance to take it past the experimental stage. Design and research are materially and financially costly.[12]

REASONS THAT THORIUM BREEDER REACTORS ARE NOT BEING BUILT

Potentially, thorium breeder reactors would enable a process of converting all the 98.3 percent of the natural uranium into radioactive substances which can maintain a sustained fission process in a chain reaction.

No one is doing this yet.

One experimental thorium breeder reactor[13] exists in Kalpakkam in India, which is also the only place where all three fissile types – U-235, Pu-239, and Th-233 – are burned.

You can read a lot about the bright future of plutonium-seeded thorium breeder reactors on the internet.[14] Conventional nuclear power stations now only use about 0.75 percent of U-235 and increase the radioactivity of what is left in the form of terrifyingly lethal contaminants known generally as the actinides. The problem of safe transport of fuels and waste that is presented in conventional nuclear power stations is likely to remain for as long as these power stations are productive. Breeder reactors would generate similar poisonous substances, but they would also burn them up in a closed and remotely controlled cycle, in continuous production of lower-grade materials which burn usefully for nuclear energy production. The waste problem would be considerably reduced even though some less long-lived wastes would still pose a storage problem.

So why aren't they being built all over the place?

The potential of thorium breeder reactors is still unproven beyond the small experimental facility in India. There is concern that *proving and building them would be very expensive*. There are still *fears that they may never work properly as units*.

There is, however, growing *support for a new paradigm of quasi-continuous self-renewing fission energy which would "eliminate" dangerous wastes*. Against this ideal is the contention that *breeder reactors could still be used to create weapons-grade plutonium*. The rebuttal of this is that weapons plutonium requires enormously more expensive Separative Work Units and is not useful or necessary for generating power, and that fast breeder reactors designed for power production would not lend themselves easily to this use. If weapons plutonium were wanted then weapons plutonium specific reactors much more suited to the task would be built.

Another reason you will read is that there is a *looming shortage of plutonium*.[15] Although when Russia and the US agreed to eliminate many of their nuclear weapons this made a lot of enriched uranium and plutonium available, much of it was snapped up by conventional reactors and nuclear submarines. Warheads became fuel for US atomic power stations. The weapons-grade plutonium is diluted to become non-explosive. Recycling it saves time and energy normally used for the enrichment process.

Another school says that *no one influential is likely to want to disturb the uranium investment market because it is so profitable, particularly in the light of impending petroleum and other fossil fuel depletion*. This factor, coupled with the legal and other set-up costs of fast breeder reactors, makes sticking to conventional reactors and mixing weapons plutonium with yellowcake more economically viable in the short to medium term – pending running out of uranium and recycled waste, and possibly pending a perfected thorium breeder reactor.

Importantly, the reliability of conventional reactors with their established safety and legal frameworks and the *comparative low cost of building new ones according to "tried and true" models discourages investment in new designs of which the setting up would entail complex and fraught negotiation of new safety and legal frameworks.*

The chief beneficiaries have *a vested interest in maintaining an industry that reprocesses and sells spent uranium* to countries which have conventional reactors but which do not have global approval to reprocess their own waste.

URANIUM SUPPLY

If uranium fueled nuclear were to expand from the 16 percent of world energy it currently supplies, then diverse projections see uranium failing to meet demand by around 2040. With no nuclear expansion, at current use it might last into the beginning of the twenty-second century.[16]

Still others have *argued that there is too much uranium around to worry about thorium or other stuff.*

SOME MORE TECHNICAL PROBLEMS WITH THORIUM AND FAST BREEDER REACTORS

The costs of developing nuclear power using thorium as fuel are increased by the engineering problems associated with the production, recycling, and containment of extremely radioactive isotopes. Far more shielding would be required than for plants currently operating, including mixed-oxide uranium and plutonium fuel, or MOX, plants, which use recycled uranium mixed with plutonium.

The thorium cycle includes the need to come to terms with exotic old and new artificial substances of extreme radioactivity. The substances include U-233, which is chemically separated from the irradiated thorium fuel, and always contains traces of uranium-232

(U-232). U-232 itself has a 69-year half life but strong gamma emitting daughter products, including thallium-208 which has a very short half life. Recycled thorium itself contains alpha emitter thorium-228 (Th-228), with a two-year half life.

The weapons proliferation risk associated with thorium FBRs is partly based on fears that U-233 might be separated on its own. The reprocessing of thorium itself is still highly experimental.[17]

The technical problems associated with the commercial development of thorium breeder reactors are so formidable, even on the scale of research possible in a country as large as India, that India could just drop its pursuit of thorium FBRs if it could obtain ready access to traded uranium.

SOME POLITICAL AND COMMERCIAL COMPLICATIONS: INDIA AS THE NEW FBR LAB

There are currently delicate international negotiations proceeding with India, which offers a huge commercial market for uranium but has an interest in developing nuclear self-sufficiency based on its huge thorium reserves. India's nuclear technology has developed independently due to being isolated through India's having developed nuclear weapons too late (1974) for inclusion as an official Nuclear Weapons State under the Nuclear Non-Proliferation Treaty (NPT).

The NPT of 1970 accorded five countries – France, China, Russia, the United Kingdom, and the United States – the exclusive official status of Nuclear Weapons States based on their having reached that status prior to 1970. Of those five countries, all but the US reprocess spent nuclear fuel. As well as having been excluded from this club, India has enduring differences with the NPT's strategies for lowering risk. Historically it has preferred to support a global policy of universal disarmament initiatives. It claims to be very uneasy about China's capabilities and not to be reassured by Pakistan's expressions of potential support for the NPT.

India has thus proceeded in comparative isolation with a civil nuclear power program, planned from the 1950s, receiving little or no fuel or technological assistance from other countries.

Up through the late 1990s India's nuclear power plants performed poorly with only 60 percent capacity.

The dot-com revolution of the 1990s saw a huge flow of Indian students and scientists into US universities, institutions, and firms. With the dot-com crash, many of them returned to India, bringing substantial technical knowledge with them. The scientific and technical community in India became very attractive for the global outsourcing of new scientific and technical developments. It is perhaps partly because of these social changes that capacity of its nuclear power plants improved markedly by 2001–02 to 85 percent.[18]

As early as the 1950s India planned for a three-stage nuclear development program. Stage One was for U-238 to be used in pressurized heavy water reactors (PHWRs). In Stage Two, the plutonium generated by these PHWRs was to be deployed to run FBRs. This has so far only been done in a small 13 MW experimental FBR at Kalpakkam. The planned FBRs were to use the plutonium mixed in a 70 percent oxide (MOX fuel) in its core within a fertile "blanket"[19] of U-233 and Th-232 which would be there to make the fuel in the core sustain fission. In Stage Three it was intended that the FBRs use Th-232 to produce U-233 as fuel for the third-stage reactors.[20] India currently has twelve nuclear power plants. The Department of Atomic Energy has government clearance to set up a 500 MW prototype of the "next-generation" FBR at Kalpakkam, with the intention of commercially exploiting thorium for its major fuel supply.

After Australia, India possesses the world's largest reserves of thorium. Use of Indian thorium would make India independent of imported uranium, including reprocessed spent uranium.

On December 9, 2006, US Congress passed the United States–India Peaceful Atomic Energy Cooperation Act, allowing shipments of nuclear fuel and technology to India for use in its civilian nuclear power program.[21] India had not yet ratified this agreement. A major point of difference was US insistence that used fuel from any US-supplied reactor must not be reprocessed.[22] This would inhibit practices in India's energy and weapons system, for both kinds of facility were, at the time of writing, still producing plutonium for reuse. The agreement would require complete separation of power facilities from weapons facilities, which were still exchanging reprocessed materials.

> The opposition to accepting safeguards on the grounds that it is difficult to separate civilian and military facilities, and that it compromises on national security, is, however, ill-founded. Demarcation of facilities as military should not be difficult but a detailed exercise of identifying these has to be carried out. The manner in which the Department of Atomic Energy (DAE) declared the Bhabha Atomic Research Centre (BARC) and a few other facilities out of bounds for AERB inspections with a single bureaucratic order in 2000,

would suggest that the process should not pose any administrative problems either. In any case, the agreement is for a *phased* declaration. But there will be a substantial cost involved and that is the price one has to pay for failing to plan for long-term fuel needs properly.

Since the research reactors Dhruva and Cirus are the chief sources of weapons-grade plutonium, and it makes no sense to use reactor-grade plutonium for weapons, one can easily demarcate all the power plants as civilian. It would seem that the main costs would pertain to replicating reprocessing plants specifically for weapon purposes because one cannot declare the existing plants – which currently reprocess spent fuel from power reactors as well as research reactors to yield plutonium for the breeder programme and weapons respectively – as military.

It is obvious that one-way traffic of nuclear material from military to civilian reactors does not pose any problem; it is only when there is a two-way traffic, as in a reprocessing plant, a dedicated facility for each objective becomes necessary because of safeguards on the material that comes in and goes out. There could be other costs involved in duplicating personnel and equipment required in this as well as other operations where people and equipment double up for the twin objectives at present.[23]

Since the UK and France, both countries which reprocess fuel, have also shown interest in the huge commercial market which India could represent, it seems likely that the pressure on the US to relent on its anti-reprocessing stance will grow. Given the profit issues and that the corporate forces have an interest in this stance changing, resistance will be difficult.

In addition, however, to purchase uranium from the 45-member Nuclear Suppliers Group would require India to sign the NPT, which India does not want to sign. It may be that the very factors which proponents of FBRs cite as discouraging their research and production in countries like the US are positives for FBR research and production in India. In this case, India is likely to be the setting for any FBR technology and production breakthrough.

COMPARING NUCLEAR TO OTHER SOURCES OF POWER WITHIN A CHANGING PARADIGM

Conventional "modern" economics that use gross domestic product (GDP) as their primary measure don't factor in many of the energy and material costs that go into GDP because those costs have been assigned no monetary value, or they have been "externalized."

The environment pays for them and that is treated as a completely free service. Although some lip-service is paid to the idea that the environment might one day stop providing these services reliably, no mechanism exists to flag this eventuality. Even where an environmental crisis is in train, attending to the problem will always show up as gains in gross domestic product (GDP).

Economic rationalism inherently relies on a paradigm of "progress" established through the exploitation of the New World,[24] followed by windfalls of coal and then oil that brought about the industrial revolution. Many hopes riding on nuclear power are based on an expectation that progress is a reliable ongoing phenomenon rather than on any real indication that nuclear will supply us when petroleum and coal cannot. Whilst overall progress in math may have truly survived the decline and fall of many complex societies, math without materials is probably just that: math.

Nuclear power has an unusual history and some special costs and risks, but so do the coal and petroleum industries. Some views that nuclear will fail the promises which currently accompany it on the market are based on assertions about materials and energy costs being unsustainable in the medium to long term. But this is also true for most conventional power sources.[25] When evaluating nuclear it is hard not to get caught up with the peculiarities of the industry. A way to mitigate this problem is to divide nuclear technology up into categories which can also be located in other technologies.

Output: energy density and reliability

The reliability and productivity of differently fueled power sources is an important factor in comparing like with like. Hydro, nuclear, and fossil fuels are much denser and more "reliable" (in the sense of continuous delivery) than wind and sun. Wind and sun, however, are more sustainable.

Nuclear power produces electricity and should be compared to other fuels which produce electricity, which are not directly comparable to the petroleum-based liquid fuels, which can be used in processes where electricity cannot – notably in road and air transport. Nuclear can, however, fuel large ships, and so can coal and wind.

Material costs

The major material costs of nuclear power plants are those of the initial separation of U-238 from yellowcake; the centrifugal separation of U-235 from U-238 (much reduced – as we know – from gaseous diffusion separation); costs of plant construction, costs of mining, and cost of electricity to run plant. These are comparable to the costs of other large power plants requiring mining, such as coal.

Specific engineering costs

Cost of engineering design and manufacture to solve complex problems grows with the necessity of innovative new design (potential thorium breeder reactors) but is comparable to those of fossil fuel plants whilst old designs are used and conventional fuel is used. Huge specific engineering costs are also involved in processing coal due to the immediate dangers of pollution as well as the dangers of the greenhouse effect.[26]

Spatial and logistical impacts

The environmental costs in terms of space and disruption to natural processes of coal mining are enormous because of the volume of coal required to produce power compared to the volume of uranium required to produce equivalent power. Likewise, the damming of rivers inundates vast areas, and impacts massively on natural systems, land tenure and previous economic development patterns due to severe interference with water flow and land use.[27] A ballpark estimate of 1 sq. mile per 1,000 MW for coal and nuclear, and 1 sq. mile per 100 MW for hydro was suggested to me to evoke the amount of land covered by plant.[28] The Three Gorges Dam, which will come at a loss of the soil in an area of 632 sq. km, will only yield 28.8 MW per sq. km.[29] This shows that power-plant size gives few clues as to what goes in and what comes out.

Uranium mining expansion

Massive expansion of uranium mines, as with coal, threatens to take place at unacceptable levels in a sort of twenty-first-century uranium and coal rush. For instance, in Australia, at Roxby Downes Olympic Dam uranium and copper mine, BHP plans to expand to possibly the largest open-cut mine operation ever, requiring huge amounts of energy and water daily,[30] the latter drawn from fossil aquifers in the Australian desert which furnish oases of enormous biological significance, underpinning the survival of entire ecologies.[31]

Cost of historic development allied to weapons

Nuclear power's historic and in some cases continuing relationship to weapons of war and war pollution is a costly association for the industry.

Special costs and risks

Some of the special costs and risks associated with nuclear power are truly catastrophic contamination in the risk of nuclear weapons proliferation, contamination of war sites and the risk of an atomic holocaust – warlike or accidental. The small amounts needed of plutonium or other fissile substances for weaponry raise the possibility of individuals being able to wield power that was previously only available to rich nations.

Bureaucratic costs

More than with "conventional" power sources, insurance has focused on the possibility of accident and on the potential severity, refusing to insure nuclear power plants, leading to a situation where governments must underwrite and oversee nuclear power.

Getting plants passed for construction and operation involves huge associated legal costs. Some call these unnecessary, but they are sociologically compelled.

The security risks and weapons association, as well as the opportunities for useful research, cause huge engagement of national and international legal resources and personnel in the difficult and expensive task of international surveillance and inspection.

A range of complex engineering problems is associated with the disposal of dangerous wastes, the shielding of personnel from radioactive materials and their isolation from the wider public, terrorism and natural disasters.

Wastes

Wastes associated with producing nuclear power seem to be relatively well monitored and their disposal is via recycling as fuel, geological sequestration (burial), or via dilution. Burying concentrated wastes is problematic because of the uncontrollability of geological events or undetected vulnerabilities which can lead to unforeseen release of concentrated toxic materials into water or air.

Before different radioactive wastes dilute (in water or air) they go through various stages of concentration, many of which could not realistically be completely monitored or controlled.

Depleted uranium weapons

The use of depleted uranium (DU) weapons has caused situations in Iraq and Kosovo where radioactivity and toxicity have obviously exceeded safe limits and the immediate boundaries in time and space of battles.[32] It seems literally insane to use DU warheads where other materials will do the same "job" with no nuclear contamination. The "collateral" of persistent postwar radiation impact on civilian, allied, and enemy troops seems self-defeating in military terms.

Tailings and radon gas emissions

These present the most serious contamination problems associated with the nuclear industry and they also occur naturally. (Only recently was it realized that radon gas occurring naturally in soils can accumulate in unventilated cellars.) The problem of radon from tailings has been very poorly handled in the recent past, with the worst kind of negligence towards miners. In the US, housing estates have been built with contaminated materials. In Australia, the long-term management of tailings around uranium mines is still not finalized. Currently, leaching into groundwater systems is practiced in the belief that dilution will take care of the matter. Fortunately, these mines are located in hot deserts with few (but not zero) inhabitants.[33]

Carbon dioxide emissions

It does seem that where separation is conducted using centrifugal technology, nuclear power plants produce less carbon dioxide (CO_2) than coal and gas power and the liquid fuels which are used for road and air transport, although these are not competitors for the same areas. The electricity to drive machinery in nuclear power plants would need to come from sources like hydro, geothermal, or nuclear to minimize CO_2 pollution.

Material sustainability

Material sustainability is relevant for comparison with alternative technologies which rely on flow and potential energies like wind, sun, water, and geothermal, which are renewable. Concerns about uranium supply and the commercial scaling up of FBRs were mentioned earlier. Commercial prospects of recovery from seawater should perhaps be compared to those for recovering marine gold.[34]

Costs of experimental plants

What are the costs of research and development in the nuclear power industry compared to those of the coal power industry? Both have huge legal and safety costs. Coal power is facing the complex logistics and theory of CO_2 mitigation schemes and ever more complex filtering systems.

Historical cost of development

Sometimes the cost of development is implicated in costing efficiency for the nuclear power industry. Aroman and Crouzet convincingly represent the French nuclear industry and its infrastructure as drawing on a debt which will cost the French economy more than it can possibly earn.[35]

These costs, if we are going to take them into account, should be compared with the costs of developing fossil fueled electricity. It could be argued then, however, that fossil fueled power has paid for its own development. This is one of the cruxes of the whole evaluation of "alternative" fuels. Arguably, coal and petroleum were their own auspicers. Fossil fuel also paid for nuclear power development, for without those riches no society would be capable of operating at the level of energy and technological sophistication required to develop or need nuclear power.

However, even the fossil fuel technological revolution stood on the shoulders of wood, and wood stood on the shoulders of soil.

In an *Oil Drum* article titled "That Cubic Mile," world use of oil in one year was estimated at one cubic mile's worth, roughly equivalent to 52 nuclear power plants each year for 50 years, or 23,850 wind turbines, or 91,250,000 solar panels, or 104 coal-fired plants, or four Three Gorges Dams every year.[36] Of course, those plants all supply electricity, not petroleum. How much more electricity would be needed to run cars on batteries, compressed hydrogen, or compressed air, in addition to the amount already used to supply heating and other daily needs for our urbanized civilization?

The nuclear plants and their supporting infrastructure would all have to be built very quickly, using the remaining abundance of petroleum and coal at a time when competition for these things is increasing.[37]

Although the world may consume around a cubic mile or more of oil this year, much of this does not translate into purposeful, quality, or socially valuable production. All powering processes are wasteful and much of the energy in oil never powers anything we want it to do.

But the world does use around one cubic mile of oil or more per annum, and the real point is, what does it use it for? And, if we didn't have all that oil, what would we choose to do without? And could we replace all that oil – or the things it does for us better than other fuel sources – with nuclear power? The fact is that the world uses all the fuel it has to do all the things it does (some which we like and others which we don't like so much). One of the things made possible by all that oil is the surplus energy to provide a hugely populous complex society with the civil and material infrastructure – for example, education systems, mining, cities, transport – that made designing and building a nuclear power station possible in the first place and also created the "need" for its energy.

NOTES

1. However, this theory is still only partial and there are many challenges to it; see, for example, http://mb-soft.com/public2/nuclei6.html.
2. Joseph F. Alward, *The Atomic Nucleus*, chapter 14, Department of Physics, University of the Pacific, http://sol.sci.uop.edu/~jfalward/physics17/chapter14/chapter14.html.
3. These may also may be categorized as pressurized heavy water reactors (PHWRs)
4. "Heavy Water Production," Federation of American Scientists, www.fas.org/nuke/intro/nuke/heavy.htm.
5. www.cns-snc.ca/Bulletin/A_Miller_Heavy_Water.pdf.
6. Jeremy Whitlock, Canadian Nuclear FAQ, www.nuclearfaq.ca/cnf_sectionA.htm.
7. "Manhattan Project," Microsoft® Encarta® Online Encyclopedia 2007, http://encarta.msn.com © 1997–2007 Microsoft Corporation.
8. "Energy Analysis of Power Systems," UIC Nuclear Issues Briefing Paper No. 57, March 2006, www.uic.com.au/nip57.htm.
9. It is very difficult to form a picture of the costs. One of many sources is http://www.uic.com.au/nip33.htm. Another paper from the same site has a good literature review and several comparative reports – see "Energy Analysis of Power Systems," UIC Nuclear Issues Briefing Paper No. 57, www.uic.com.au/nip57.htm.
10. A report by the ISA of the University of Sydney, "Life Cycle Energy Balance and Greenhouse Gas Emissions of Nuclear Energy in Australia," includes a full cycle energy analysis. Their spreadsheet is available at www.peakoil.org.au/isa.nuclear-calculator.xls.
11. Ibid.
12. www.uic.com.au/nip98.htm.
13. Also known as a thermal breeder reactor and an advanced heavy water reactor.
14. See, for instance, Michael Anissimov's "A Nuclear Reactor in Every Home," October 16, 2006, www.acceleratingfuture.com/michael/blog/?p=212.
15. The Megatons to Megawatts Program is the name given to the program that implemented the 1993 United States–Russia non-proliferation agreement to convert high-enriched uranium (HEU) taken from dismantled Russian nuclear weapons into low-enriched uranium (LEU) for nuclear fuel. From 1995 through to mid-2005, 250 metric tonnes of high-enriched uranium (enough for 10,000 warheads) were recycled into low-enriched uranium. The goal is to recycle 500 metric tonnes by

2013. Much of this fuel has already been used in many nuclear power plants in the US, as it is indistinguishable from normal fuel, www.usec.com/v2001_02/HTML/megatons.asp.

16. Figure 6, "History and forecast of uranium production," in "Uranium Depletion and Nuclear Power: Are We at Peak Uranium?" www.theoildrum.com/node/2379, accessed December 3, 2007. See also the World Nuclear Association Position Statement, cited in "Can Uranium Supplies Sustain the Global Nuclear Renaissance?": "According to the Authoritative 'Red Book' produced jointly by the OECD's Nuclear Energy Agency and the UN's International Atomic Energy Agency, the world's present known economic resources of uranium, exploitable at below $80 per kilogram of uranium, are some 3.5 million tonnes. This amount is therefore enough to last for 50 years at today's rate of usage – a figure higher than for many widely used metals" (www.uic.com.au/WNA-UraniumSustainability.pdf). Another source is Jan Willem Storm van Leeuwen and Philip Smith, "Rebuttal of World Nuclear Association Critique of the Analysis 'Can Nuclear Power Provide Energy for the Future; Would it Solve the CO_2-emission Problem?'" www.greatchange.org/bb-thermochemical-rebuttal_WNA.html.
17. Correspondence with Ian Hore-Lacy, Director – Information, Australian Uranium Association, Melbourne, www.uic.com.au, April 30, 2007, and May 1, 2007.
18. www.uic.com.au/nip45.htm.
19. The core is the central part of a nuclear reactor containing the fuel elements and any moderator. A fast neutron reactor is configured to produce more fissile material than it consumes, using fertile material such as depleted uranium in a blanket around the core (definitions from www.world-nuclear.org/info/inf51.htm).
20. *Frontline*, vol. 22, issue 16 (July 30–August 12, 2005), http://tinyurl.com/362kjb.
21. "Nuclear Technology Milestones 1942 to Present," Nuclear Energy Institute, Washington, www.nei.org/index.asp?catnum=3&catid=265.
22. "Sticking Points in US–India Talks," Reuters, March 30, 2007; *Nucleonics Week*, April 12, 2007; *Financial Times*, April 4, 2007.
23. R. Ramachandran, "Behind the Bargain," *Frontline*, vol. 22, issue 16 (July 30–August 12, 2005), www.hinduonnet.com/fline/fl2216/stories/20050812005700700.htm.
24. Elegantly demonstrated in R.E. Dunlap and W.R. Catton, Jr., "A New Ecological Paradigm for a Post-Exhuberant Sociology," *American Behavioral Scientist*, vol. 24, no. 1 (September/October 1980), Sage Publications Inc., pp. 15–47.
25. For a short review of the literature associated with studies done on this see "Energy Analysis of Power Systems," UIC Nuclear Issues Briefing Paper No. 57, www.uic.com.au/nip57.htm.
26. Barbara Freese, *Coal: A Human History*, Arrow Books, UK, 2006, chapter 10.
27. For more about the problems with dams see documents at Dialogue on Dams and Rivers, The Aspen Institute, http://tinyurl.com/32p6pw.

28. Jeff Barton, energyresources@yahoogroups.com, Tuesday, May 15, 2007, 2.59 p.m., subject "RE: [energyresources] comparing size of plant for Coal, nuclear, & hydro."
29. "When completed, Three Gorges Dam will be the world's largest hydropower facility with a generation capacity of 18,200 megawatts. It will simultaneously supply flood storage and enhance navigation along the Yangtze River. The structure will create a reservoir more than 600 kilometers long and 1,100 meters wide, capable of storing 39.3 billion cubic meters of water. Construction of the dam, which began in 1993, requires the inundation of 632 square kilometers of existing land and will cause the permanent relocation of over 1.2 million people" (www.waterencyclopedia.com/Da-En/Dams.html).
30. James Finch, "Forecasted Uranium Production Not Always a 'Sure Thing'," July 18, 2007, www.stockinterview.com/News/07182007/Uranium-production-uncertainty.html.
31. "Greens Want to End Olympic Dam's Special Treatment," www.abc.net.au/news/stories/2007/06/06/1943630.htm, June 6, 2007. See also "End Olympic Dam Special Treatment," http://indyhack.blogspot.com/2007/06/end-olympic-dam-special-treatment.html, and "Uranium Mining in Arkaroola," www.markparnell.org.au/campaign.php?campaignn=19.
32. D. Bradbury and P. Scott, *Blowin' in the Wind*, 2005 (Film), www.bsharp.net.au/htm/the-film.htm.
33. "Review of Environmental Impacts of Acid in-situ Leach Uranium Mining Process," CSIRO Land & Water Report, 2004 (commissioned by SA Parliament to address issues at Beverley and Honeymoon), www.epa.sa.gov.au/pdfs/isl_review.pdf, "Uranium Mill Tailings."
34. Sugo; Takanobu (Gunma-ken, JP); Katakai; Akio (Gunma-ken, JP); Seko; Noriaki (Gunma-ken, JP) (Inventors), "Collector for adsorptive recovery of dissolved metal from sea water," US Patent No. 6863812, www.uspatentserver.com/686/6863812.html.
35. Marc Saint Aroman and André Crouzet, "French Nuclear Power and the Global Market: An Economic Illusion," *The Final Energy Crisis*, Pluto Press, London, 2005.
36. Engineer-Poet (article) and 2007 Brian Christie (illustration), "That Cubic Mile," *The Oil Drum*, February 28, 2007, www.theoildrum.com/node/2320.
37. On prohibitive marginal costs of research and development, see J. Tainter, *The Collapse of Complex Societies*, Cambridge University Press, Cambridge, 1988; 12th reprint, 2005, p. 212.

18
Fusion Illusions
Fundamental Questions on the Viability of Commercial Energy Production from Deuterium-tritium Fusion

Michael Dittmar

INTRODUCTION

A belief that controlled nuclear fusion will solve all future energy problems is widespread within the scientific community and especially among physicists. Supporting this view are the ideas that the raw materials (hydrogen and deuterium) exist in almost infinite quantities because they can be extracted from ordinary water; the process is "clean" because no long-lived radioactive heavy elements are produced; and the process is carbon dioxide (CO_2) free. The last point was added only a few years ago.

We are told that only one "minor" hurdle remains: How to keep the raw materials under high pressure, at about 100 million degrees Celsius, for a long time, as a "plasma"?[1]

Using enormous amounts of electric energy to create sophisticated magnetic fields, fusion scientists try to achieve the fusion conditions which occur naturally in the center of the sun. Enormous and impressive progress on magnetically confined plasmas at high temperatures has been achieved during the last 50 years and the fusion of deuterium and tritium to helium plus "energy" was successfully demonstrated in experiments some ten years ago.

The experiments in question achieved a thermal power production of 4 million watts (or 4 megawatts, MW) for five seconds and 15 MW for about one second. The so-called Q-value, the ratio of produced energy to the input energy, was still below 1. That is, the experiments used more energy than was liberated in the process. Yet these experiments are considered proof that the technological problems of fusion can be resolved in principle, and that a worldwide cooperative effort plus more money will overcome the remaining practical difficulties in the next fusion project; that today's dreams about "infinite cheap and clean energy" will then become reality. This is the story physics students learn at universities and later tell friends, school children, journalists, and politicians.

More advanced courses offer more detail, such as the fact that the deuterium-tritium fusion reaction, $D + T \rightarrow He^4_2 + n + 18$ MeV (million electron volts), has the lowest required temperature combined with the highest fusion reaction rate.[2] For this reason it is the only reaction that is envisaged for a future commercial fusion reactor.

The 18 MeV of liberated energy would be shared between the neutron with a kinetic energy of 14 MeV and the helium nucleus (He^4_2) with 4 MeV. Escaping the reactor core, the neutron would heat up a surrounding medium. The 4 MeV helium nucleus remains in the reactor core and its energy is supposed to keep the plasma temperature high enough for further fusion reactions.

These arguments seem to be sufficient to convince policy makers and the public in general to provide the required funding. Although successfully used to elicit the €10 billion (US$14.6 billion) funding for the International Thermonuclear Experimental Reactor, or ITER,[3] as the next step toward a commercial fusion reactor, the whole story is a great deal more complex.

The next section in this chapter, "Energy from Nuclear Fusion," describes the well known physics of nuclear fusion, outlining the path to a commercial fusion test reactor as envisaged and explained by the proponents of the ITER project. Following this, "Remaining Barriers to Fusion Energy on Planet Earth" chronicles the achievements of the last 50 years of intense and expensive research, evoking the barriers remaining before we can even begin to think about designing a commercial fusion test reactor. "Illusions of Tritium Self-sufficiency" then explains why a real fusion reactor would have to achieve self-sufficient tritium production and why it seems that the fulfillment of the necessary breeding parameters would be miraculous.

The chapter concludes that to date not a single experiment has been able to demonstrate that the theoretically imagined process can function in practice. In addition, the computer-simulated "experiments" discussed do not even include a mechanism on how any tritium produced could be extracted. Curiosity is expressed as to why this apparently crucial problem related to tritium self-sufficiency does not appear to arouse concern within the fusion program community. In fact the fusion community generally seems to avoid discussing this problem in a transparent manner, if at all.

Questions therefore remain inadequately addressed and a forum for related serious interdisciplinary scientific debate does not exist.

For these reasons, chapters like this one may be needed to draw attention to and stimulate public discussion of scientific concern about the ITER project's approach to solving the energy problem.

ENERGY FROM NUCLEAR FUSION

Over 100 years ago, physicists began to understand how energy production in stars could last for billions of years. Our own sun, over the 5 billion years of its existence, has emitted energy of about 4×10^{26} joules per second, corresponding to the fusion of about 600 million tonnes of hydrogen into helium per second. This process may continue for another 5 billion years, since enough hydrogen exists.

The dominant fusion cycle in the sun starts with the fusion of two hydrogen nuclei – protons – since hydrogen has no neutrons and only one electron. This process results in the heavy hydrogen isotope called deuterium, a proton-neutron bound state.[4] Next, another proton is added to make He^3, the helium nucleus made of two protons and one neutron. Finally, two He^3 nuclei are fused into He^4, plus two protons and a large amount of energy. The enormous gravitational pressure in the sun leads to a high enough density and temperature at its center such that continuous fusion occurs.

The production of only 1 kg of helium "liberates" the heat energy of almost 10^9 kilowatt hours (kWh) and scientists quickly saw the energy potential of nuclear processes, especially with the discovery of the neutron in 1931. The neutron-induced nuclear fission of uranium into two lighter elements, plus two to three neutrons and a large amount of energy, were observed in experiments at the end of 1938. Only four years later, on December 2, 1942, a nuclear chain reaction with a power of 0.5 watts (W) (and up to 200 W a later time) was sustained by E. Fermi and his team below the Chicago University football stadium.[5]

The next steps in using nuclear energy were the explosions of the Hiroshima and Nagasaki fission bombs, on August 6 and 9, 1945, resulting in more than 100,000 human deaths. The nuclear arms race followed during the Cold War, and led only a few years later to US and Soviet hydrogen fusion bombs. These bombs were up to 1,000 times more powerful than the Hiroshima fission bomb. The largest fusion bomb to date was exploded in 1961 by the Soviet Union.

It turned out to be relatively straightforward to operate a controlled fission chain reaction, where the "liberated" heat could be used to operate a generator and produce electricity.

For example, the first commercial fission reactor was producing electricity in 1954, only nine years after the first fission bomb was successfully tested. Today, about 440 commercial fission reactors, with a total electric power of about 370,000 MW, operate in 31 countries. They produce about 16 percent of the world's electric energy and thus roughly 2.5 percent of today's world energy use.

Whatever one thinks of the wisdom of nuclear fission power, the quick scientific and technical success in bringing this form of power online gave rise to a euphoric belief in continuous scientific and technological progress. Thus, after World War II, many nuclear pioneers expected that nuclear fusion would provide their grandchildren with cheap, clean, and essentially unlimited energy. But obviously this has not happened, and today's fusion optimists still do not expect the first commercial fusion reactor prototype for at least another 50 years.

Generations of physicists and physics teachers have, however, been taught at university, and have gone on to teach others, (a) that progress made in fusion research is impressive, (b) that controlled fusion is probably only a few decades away, and (c) that – given sufficient public funding – no major obstacles stand between us and success in this field. Quotes from physics textbooks often reflect this sort of optimism:

> The goal seems to be visible now ...[6]

> It will most likely take until the year 2000 to bring a laboratory reactor to full commercial utilization ...[7]

> As the construction of a fusion reactor implies a large number of unsolved practical problems, one can not expect that fusion will become a usable energy resource during some decades! Within a longer time scale however it seems possible![8]

A similarly uncritical media waxed enthusiastically about the recent decision by the "world's leaders" to provide the €10 billion needed to start the ITER fusion project. In the ITER outreach pages one could read how this funding would permit worldwide collaboration of scientists whose combined effort in nuclear fusion would solve potential future energy problems.

Such technological conviction finds a receptive audience in a public increasingly concerned about possible collateral damage from fossil fueled "global warming," the problems of global injustice and related poverty, real and imagined fears about nuclear fission energy, and, of course, the prospect of rising energy costs.

So the public welcomes the tacit message that *our fusion scientists, engineers and politicians are working round the clock to make sure that we have a solution before we run out of fossil fuels or global warming boils us all.*

ITER's main arguments, summarized in their outreach program are:

- Understanding of plasma physics has progressed so impressively that the production of plasma temperatures and density is near to the so called "breakeven" point. The "breakeven" point means that more energy can be extracted than is needed to heat the plasma to the incredible temperatures required for fusion. Much of ITER's language is more metaphorical than technical, for example, "bringing the sun to the earth." Sometimes it is acknowledged that we are only talking about the deuterium-tritium fusion reaction and that all other fusion reactions and especially the one in the sun are out of reach.
- The necessary "fusion fuels" are the hydrogen isotopes, deuterium (a bound state of one proton and one neutron), and tritium (one proton and two neutrons). The deuterium isotope is found in ordinary water with a concentration of about 0.02 percent per hydrogen atom. Tritium is an unstable isotope with a half life of 12.3 years. A real fusion reactor will need initial seeding with tritium from an external source before it can provide its own self-sustained tritium breeding chain with a capacity of at least 55.6 kg per year and a thermal power production of 1,000 MW production per year. The tritium breeding should take place within a lithium "carpet" surrounding the plasma zone. The plasma zone has a volume of more than 1,000 cubic meters (m^3), and the carpet should have a thickness of at least one meter. The self-sufficient fusion chain is then imagined in these steps:
  (a) The fusion process $D + T \rightarrow He^4_2 + n +$ energy. About 80 percent of the liberated fusion energy is taken up by the

neutron which leaves the plasma zone. The energy carried by the helium nucleus will be used to keep the plasma hot enough for subsequent fusion reactions. After the neutron has been slowed down, in a process similar to the neutron moderation in a fission reactor, the breeding reaction $n + Li^6_3 \rightarrow He^4_2$ (tritium) should take place. Both deuterium and lithium are relatively cheap and exist for all practical purposes in essentially unlimited quantities on our planet.

(b) The tritium produced in this process must then be extracted from the thick carpet material and collected with negligible losses. The tritium must then be transferred back to the reactor center.

- Fusion optimists claim that if the planned ITER experiments function as expected today, then the first real electricity production will be possible in a later, much bigger fusion reactor, called DEMO, perhaps just 30 years from now.[9] If all subsequent DEMO experiments were to function as they predict today, then in roughly 50 years from now, some commercial electricity production could be envisaged from a prototype reactor, called PROTO.

This gives the impression that our human institutions are capable of planning over much longer timescales than the four-year lifetimes of governments. And it seems that if scientists get sufficient funding, they can solve, within 50 years or so, all the potential energy problems feared by the so-called energy pessimists. Who then could doubt that the world's best scientists and engineers will be able to build upon and dramatically extend the impressive technological progress of the last 50–100 years? It seems, in short, that all the necessary planning for our energy future is in hand and that it would be foolish to listen to rumors about a looming world energy crisis.

Unfortunately, this essentially is the same story we have been hearing for some 30–50 years now. The most significant difference in today's version is that controlled fusion is now seen to be about 50 years away instead of 20–30 years away. Yet the ITER scientists never seem to challenge any of the above assumptions and work enthusiastically on their specific assigned problems.

In what follows, this chapter challenges the assumptions of the ITER project and quantifies the arguments of fusion skeptics. It starts by comparing what fusion scientists have achieved so far, what they expect to achieve with a successful ITER program and what would actually be required for a hypothetical commercial fusion reactor. This overview of the remaining problems facing commercial fusion, many of which will not even be studied within the ITER project, will demonstrate the need for a large number of "divine interventions" if belief in controlled fusion is to be kept alive.

Chief among the problems to be discussed here is the envisaged self-sufficient tritium breeding cycle. It seems to us as unlikely as the idea of the operation of a fission reactor with heavy non-radioactive elements like lead. In fact, the points explained in this chapter demonstrate that enough knowledge has been accumulated on this subject already to safely conclude that whatever might justify the €10 billion ITER project, it is not energy research.

REMAINING BARRIERS TO FUSION ENERGY ON PLANET EARTH

Producing electricity from controlled nuclear fusion would require overcoming at least four major obstacles. The removal of each obstacle would need major scientific breakthroughs before any reasonable expectation might be formed of building a commercial prototype fusion reactor.

It should be alarming that at best only the problems about plasma control, described in point (1) below, might be investigated within the scope of the ITER project. Where and how the others might be dealt with is anyone's guess.

These are the four barriers:

1. Commercial energy production requires the achievement of steady-state fusion conditions for a deuterium-tritium plasma on a scale comparable with today's standard nuclear fission reactors with outputs of 1,000 MW electric (MWe) and about 3,000 MW thermal (MWth) power.[10] The current ITER proposal foresees a thermal power of only 400 MW using a plasma volume of 840 m^3. Originally it was planned to build ITER with a plasma volume of 2,000 m^3 corresponding to a thermal fusion power of 1,500 MW, but it was realized quickly within the fusion community that the original ITER version would never receive the required funding. Thus a smaller, much less ambitious version of the ITER project was proposed and finally accepted in 2005.

Today's 1,000 MWe fission reactors function essentially in a steady-state operation at nominal power and with an availability time over an entire year of roughly 90 percent. The deuterium-tritium fusion experiments have so far achieved short pulses of fusion power of 15 MWth for about one second and 4 MWth for about five seconds corresponding to a liberated thermal energy of about 5 kWh. The Q-value – produced energy over input energy – for both pulses was 0.65 and 0.2 respectively.

If everything works as planned, it will be 2022 (or about six years after the first plasma experiment has been performed) before ITER tries, for the first time, to achieve a power output of 500 MWth with a Q-value of up to 10 and for about 400 seconds. Compare that to the original ITER proposal which was 1,500 MWth, with a Q value between 10 and 15 and for about 10,000 seconds. ITER proponents explain that the achievement of this goal would already be an enormous success. But this goal, even if it can be achieved in 2022, pales in comparison with the requirements of steady-state operation, year after year, with only a few minor controlled interruptions.

Previous deuterium-tritium experiments used only minor quantities of tritium and yet lengthy interruptions between successive experiments were required because the radiation from the tritium decay was so excessively high. In earlier fusion experiments, such as the 1997 UK deuterium-tritium fusion project, JET,[11] the energy liberated in the short pulses came from burning (fusing) about 3 micrograms (μg) (3×10^{-6} g) of tritium, starting from a total amount of 20 g of tritium. This number should be compared with the few kilograms of tritium required to perform the experiments foreseen during the entire ITER lifetime and the still greater quantities that would be required for a commercial fusion reactor. A 400-second fusion pulse with a power of 500 MW corresponds to the burning of about 0.035 g (3.5×10^{-2} g) of tritium. A very large number when compared to 3 μg, but a tiny number when compared with the yearly burning of 55.6 kg of tritium in a commercial 1,000 MWth fusion reactor.

The achieved efficiency of the tritium burning (that is, the amount that is burned divided by the total amount that was required to achieve the fusion pulse) was roughly one part in a million in the JET experiment and is expected to be about the same in the ITER experiments, far below the efficiency required to burn 55.6 kg of tritium per year.

Moreover, in a steady-state operation the deuterium-tritium plasma will be "contaminated" with the helium nucleus it produces and some instabilities can be expected. Thus a plasma cleaning routine that would not cause noticeable interruptions of production in a commercial fusion plant is needed. ITER proponents know that even their self-defined goal (a 400-second-long deuterium-tritium fusion operation within the relatively small volume of 840 m^3) presents a great challenge. They seem surprisingly silent about the difficulties involved in reaching steady-state operation for a full-scale fusion power plant.

2. The material (called a "blanket") that surrounds and contains thousands of cubic meters of plasma (containing the active particles) in a full-scale fusion reactor has to fulfill two requirements. First, it has to survive an extremely high neutron flux (neutron radiation) with energies of 14 MeV, and second, it has to do this not for a few minutes but for many years. It has been estimated that in a full-scale fusion power plant the neutron flux (or bombardment) will be at least ten to twenty times larger than in today's state-of-the art nuclear fission power plants. Since the neutron energy is also higher, it has been estimated that – with such a neutron bombardment – each atom in the solid that surrounds the plasma will be displaced about 475 times over a period of five years.[12] Second, to further complicate matters, the material in the so called first wall (FW) around the plasma will need to be very thin, in order to minimize inelastic neutron collisions resulting in the loss of neutrons (more details next section), yet at the same time thick enough so that it can resist both normal and accidental collisions from the 100-million-degree hot plasma for years.

The "erosion" for carbon-like materials from the neutron bombardment has been estimated to be about 3 mm per "burn" year, and even for materials like tungsten it has been estimated to be about 0.1 mm per burn year.[13]

No known material can even come close to meeting the requirements just described. Exactly how a material could be designed and tested to meet those requirements remains a mystery, because tests with such extreme neutron fluxes cannot be performed either at ITER or at any other existing or planned facility.

3. Because radioactive decay of even a few grams of tritium creates radiation dangerous to living organisms, those who work with it must take sophisticated protective measures.

Tritium is, moreover, chemically identical to ordinary hydrogen and as such very active and difficult to contain. Since tritium is also a necessary ingredient in hydrogen fusion bombs, there is additional risk that it might be stolen. So, handling even the few kilograms of tritium that are foreseen for ITER is likely to create major headaches about radiation protection and nuclear weapons proliferation.

The above challenges are essentially ignored in the ITER proposal, and the only thing the radiation and weapons proliferation concern groups have to work with are design studies based on computer simulations. This may not be of concern to the majority of ITER's promoters today, since they will be retiring before the tritium problem starts in something like ten to fifteen years from now.[14] At some point, however, ITER will have to face these problems if it actually begins work on a real fusion experiment with many tens of kilograms of tritium.

4. Problems related to tritium supply and self-sufficient tritium breeding will be discussed in detail in the following section of this chapter. First, however, I will describe qualitatively two problems which it seems it would require simultaneous miracles to solve.

- The neutrons produced in the fusion reaction will be emitted essentially isotropically in all directions around the fusion zone. These neutrons must somehow be convinced to escape without further interactions through the first wall surrounding the few 1,000 m^3 plasma zone. Next, the neutrons have to interact with a "neutron multiplier" material like beryllium in such manner to increase the neutron flux without transferring too much energy to the remaining nucleons. The neutrons then must transfer their energy without being absorbed (for example, by elastic scattering[15]) to some kind of gas or liquid, like high-pressure helium gas, within the lithium carpet. This heated gas or liquid has to be collected somehow from the gigantic carpet volume and must be encouraged to flow to the outside. As in any existing power plant, this heat can be used to power a generator turbine. The gas or liquid should be as hot as possible, in order to achieve reasonable efficiency for electricity production. As we know already, however, the lithium carpet temperature can't be too hot, thus limiting possible efficiencies well below the ones from today's not very efficient nuclear fission reactors.

  Once the heat is extracted and the neutrons have slowed sufficiently, they must interact inelastically with the Li^6 isotope, which makes up about 7.5 percent of natural occurring lithium. The minimum thickness required of the so called lithium carpet that surrounds the entire plasma zone has been estimated to be at least one meter. Unfortunately, lithium, like hydrogen, in its pure form is chemically highly reactive. If used in a chemically bound state with oxygen, for example, the oxygen itself could interact and absorb neutrons, something that must be avoided. In addition, that the lithium and the tritium produced would react chemically and that some seven tritium atoms will be blocked within the carpet, has certainly not been included in any present computer modeling. Unfortunately, additional neutron and tritium losses cannot be allowed. Reasons will be described in more detail in the next section.

- Next, the engineers need to find an efficient way to extract the tritium quickly before it decays and without loss from this lithium carpet. We are talking about a huge carpet here, one that surrounds the few 1,000 m^3 plasma volume. Extracting and collecting the tritium from this huge lithium carpet will be very tricky indeed, since tritium penetrates thin walls relatively easily, and since accumulations of tritium are highly explosive.[16] And finally assuming we get that far, the extracted and collected tritium and deuterium, which both need to be extremely clean, need to be transported, without losses, back to the reactor zone.

Each of the unsolved problems described above is, by itself, serious enough to raise doubts about the envisaged success of commercial fusion reactors. But the self-sufficient tritium breeding is especially problematic, as will be described in the next section.

ILLUSIONS OF TRITIUM SELF-SUFFICIENCY

In fact, a self-sustained tritium fusion chain appears to be not simply problematic but absolutely impossible. To see why, we will now look into some details based on what is already known about this problem.

A central quantity for any fission reactor is its criticality, namely that exactly one neutron, out of the two to three neutrons "liberated" per fission reaction, will enable another nuclear fission reaction. More than 99 percent of the liberated fission energy is taken by the heavy fission products such as barium and krypton, and this energy is relatively easily transferred to a cooling medium. The energy of the neutrons produced in fission is about 1 MeV. In order to achieve criticality, the surrounding material must have a very low neutron absorption cross-section and the neutrons must be slowed down to eV energies. For a self-sustained chain reaction to happen, a large amount of uranium-235, enriched to 3–5 percent, is usually required.

Once the nominal power is obtained, the chain reaction can be regulated using materials which have a very high neutron absorption cross-section. Fast breeder reactors (FBRs) require higher enrichment than moderated reactors, and a much higher enrichment of up to 90 percent is required for bombs.

In contrast to fission reactions, only one 14 MeV neutron is liberated in the $D + T \rightarrow He + n$ fusion reaction. This neutron energy has to be transferred to a medium using elastic collisions.

Once this is done, the neutron is supposed to make an inelastic interaction with a lithium nucleus, splitting it into tritium and helium.

Starting with the above reaction one might easily calculate how much tritium burning is required to operate a continuous operating commercial fusion reactor assuming a power production of 1,000 MWth).[17] One finds that about 55.6 kg of tritium needs to be burned per year with an average thermal power of 1,000 MW.

Today's tritium is extracted from nuclear reactors at extraordinary cost – about US$30 million per kilogram from Canadian heavy water reactors. These old heavy water reactors will probably stop operation around the year 2025 and one expects that a total tritium inventory of 27 kg will be accumulated by that year.[18]

Once these reactors stop operating, however, this inventory will deplete by more than 5 percent per year due to its radioactive decay alone. Tritium has a half life of 12.3 years. These inexorable depletion rates mean that at best only about 7 kg of tritium might remain to start the prototype "PROTO" fusion reactor by 2050, which is the earliest date fusion optimists contemplate for its beginning operation. (Normal fission reactors produce at most 2–3 kg per year and extraction costs have been estimated to be about US$200 million per kg.)[19] It is thus obvious that any future fusion reactor experiment beyond ITER must not only achieve tritium self-sufficiency, it must create more tritium than it uses if there are to be any further fusion projects.

The particularly informative website of professor Abdou[20] from UCLA, one of the world's leading experts on tritium breeding, allows us to get some relevant numbers both about the basic requirements for tritium breeding and the state of the art today.

But first things first: understanding such "expert" discussions requires an acquaintance with some key terms:

- The "required tritium breeding ratio" (rTBR) stands for the minimal number of tritium nuclei which must be produced per fusion reaction in order to keep the system going. It must be larger than 1, because of tritium decay and other losses and because of the necessary inventory in the tritium processing system and the stockpile for outages and for the startup of other plants. The rTBR value depends on many system and technology parameters.
- The "achievable tritium breeding ratio" (aTBR) is the value obtained from complicated and extensive computer simulations – so-called three-dimensional simulations – of the carpet with its lithium and other materials. The aTBR value depends on many parameters like the first wall material and the incomplete coverage of the breeding carpet.
- Other important variables are used to define the quantitative value of the rTBR. These include: (1) the "tritium doubling time," the time in years required to double the original inventory; (2) the "fractional tritium burn-up" within the plasma, expected to be at best a few percent; (3) the "reserve time," the tritium inventory required in days to restart the reactor after some system malfunctioning with a related tritium loss; and (4) the ratio between the calculated and the experimentally obtained TBR.

The handling of neutrons, tritium, and also lithium requires particular care, not only because of radiation danger, but also because tritium and lithium atoms are chemically very reactive elements. Consequently, real-world, large-scale experiments are not easily performed and today's understanding about the tritium breeding is based almost entirely on complicated and extensive computer simulations, simulations which can only be done in a few places around the world.

Sawan and Abdou describe results from some of these simulated experiments.[21] The authors assume that a commercial fusion power reactor of 1,500 MWth (burning about 83 kg of tritium per year) would require a long-term inventory of 9 kg, and they further assume that the required start-up tritium is available. For a doubling time of five years and a 5 percent fractional tritium burn-up, they assume the absolute minimum required rTBR is 1.2, or perhaps a bit more.

Other numbers can be read from their figures. For example, with shorter doubling times, the rTBR would have to be still larger – for example, a one-year doubling time would demand a minimum rTBR of around 1.5. The corresponding result for the rTBR with a 1 percent fractional tritium burn-up[22] and a five-year and one-year doubling time are 1.4 and 2.6 respectively.

The importance of short tritium doubling times can be understood easily using the following calculation. Assuming these numbers can be achieved and that the 27 kg tritium (at 2025) less 9 kg (long-term inventory) would be available at start-up, then 18 kg could be burned in the first year. A doubling time of four years would thus mean that such a commercial 1,500 MWth reactor could operate at full power for only about eight years after the start-up.

Even these rTBR estimates are far too optimistic since they fail to consider a number of potential losses related to the extraction, collection and transport of the tritium.

The details become even more troubling when we turn to the tritium breeding numbers that have been obtained with today's computer simulations.

After many years of detailed studies, current simulations show that today's blanket designs have, at best, achieved TBRs of 1.15. Using this number, Sawan and Abdou conclude that theoretically a small window for tritium self-sufficiency still exists. This window requires (1) a fractional tritium burn-up of more than 5 percent, (2) a tritium reserve time of less than five days, and (3) a doubling time of more than four years. But even with these highly theoretical numbers in mind, the authors have difficulty envisaging a real operating power plant. In their words, "for fusion to be a serious contender for energy production, shorter doubling times than 5 years are needed." Indeed, doubling times much shorter than five years appear to be required. This means TBRs much higher than 1.15 are required. Interestingly, Sawan and Abdou also acknowledge that current systems of tritium handling need to be explored further. This probably means that the tritium extraction methods from nuclear fission reactors are nowhere near meeting the requirements.

Sawan and Abdou also summarize various effects which reduce the obtained aTBR numbers once more realistic reactor designs are studied and consideration given to structural materials, gaps, and first wall thickness. For example, they find that as the first wall, made of steel, is increased by 4 cm starting from a 0.4 cm wall, the aTBR drops by about 16 percent. It would be interesting to compare these assumptions about the first wall with the ones used in previous plasma physics experiments like JET and the one proposed for ITER. Unfortunately, so far I have not been able to obtain any corresponding detailed information. However, as one expects that the first wall in a real fusion reactor will erode by up to a few millimeters per fusion year, the *requisite* thin walls seem to be another impossible assumption made by the fusion proponents.

Other effects, as described in detail by Sawan and Abdou,[23] are known to reduce the aTBR even further. The most important ones come from the cooling material needed to transport the heat away from the breeding zone, from the electric insulator material, from the incomplete angular coverage of the inner plasma zone, with a volume of more than 1,000 m^3, and from the plasma control requirements.

This list of problems is already very long and, in the absence of answers from ITER, removes the basis for belief in a self-sufficient tritium. In addition, some still very idealized TBR tests have actually been performed. All these real experiments show that the measured TBR results are consistently about 15 percent lower than the modeling predicts, according to Sawan and Abdou:[24]

> the large overestimate [of the aTBR] from the calculation is alarming and implies that an intense R&D program is needed to validate and update ... our ability to accurately predict the achievable TBR.

Another interpretation might be that published results from experiments to date consistently fail to show that a window for self-sufficient tritium breeding currently exists. This suggests that extant proposals for future tritium breeding have no demonstrable scientific basis.

In the absence of any supplementary evidence, one must conclude that they rely on commercial hope, technological faith, and public misunderstanding.

FORGET DREAMS OF CONTROLLED NUCLEAR FUSION

As explained in the previous sections, there is a long list of fundamental problems with controlled fusion. Each appears to be big enough to raise serious doubts about the viability of the chosen approach to a commercial fusion reactor and thus about the €10 billion ITER project.

Those not familiar with the handling of high neutron fluxes or the possible chemical reactions of tritium and lithium atoms might suppose that these problems are well known within the fusion community and are being studied intensively. The truth is that none of these problems has been studied intensively. Even with the ITER project, the only problems that might be studied relate to some of the plasma stability issues outlined earlier in this chapter. All of the other problem areas are essentially ignored in today's discussions among "ITER experts."

Confronted with the seemingly impossible tritium self-sufficiency problem that must be solved before a commercial fusion reactor is possible, my experience is that "ITER experts" change the subject and tell you that this is not a problem for their ITER project. In their view it will not be until the next generation of experiments – experiments that will not begin for roughly another 30 years, according to official plans – that issues related to tritium self-sufficiency will have to be dealt with. Perhaps they are also comfortable with the fact that neither the problems related to material aging due to the high neutron flux nor the problems related to tritium and lithium handling can be tested with ITER. Perhaps they are expecting miracles from the next generation of experiments.

However, among those who are not part of ITER and those who are not expecting miracles, it seems that times are changing. More and more scientists are coming to the conclusion that commercial fusion reactors can never become reality. Some are even receiving a little attention from the media as they argue more and more forcefully that the entire ITER project has nothing to do with energy research.[25]

More attention should be paid to Professor Abdou. In a presentation prepared on behalf of the US fusion chamber technology community for the US Department of Energy (DOE) Office of Science on Fusion Chamber Technology, in 2003, Abdou wrote: "Tritium supply and self-sufficiency are 'Go–No Go' issues for fusion energy, [and are therefore] as critical NOW as demonstrating a burning plasma." He pointed out: "There is NOT a single experiment yet in the fusion environment that shows that the DT fusion fuel cycle is viable" (capitalization in original). "Proceeding with ITER makes Chamber Research even more critical," he stated, and asked: "What should we do to communicate this message to those who influence fusion policy outside DOE?"[26]

In short, to go ahead with ITER without addressing these chamber technology issues makes no sense at all.

In light of everything that has been written in this chapter, it seems urgent that ITER should publish its research with the appropriate conclusions to the effect that:

1. Today's achievements in all relevant areas are still many of orders of magnitude away from the basic requirements of a fusion prototype reactor.
2. No material or structure is known which can withstand the extremely high neutron flux expected under realistic deuterium-tritium fusion conditions.
3. Self-sufficient tritium breeding appears to be impossible to achieve under the conditions required to operate a commercial fusion reactor.

The taxpayers, the policy makers, and the media need to be told that after 50 years of very costly research conducted at various locations around the world, enough knowledge exists to state that commercial energy production from nuclear fusion is never going to be a reality.

ITER would need to advise without further delay that it could only continue to function as an expensive experiment to investigate some fundamental aspects of plasma physics.

As with all grand projects, defeat is hard to acknowledge, but it is time to review ITER's funding on equal terms with all other fundamental research projects. ITER's wealth of material and human skills resources should be transferred to deal with our emerging energy crisis in a realistic way.

These realities need to be addressed at a scientific level before they become a real political problem.

NOTES

1. Plasma is an electrically neutral medium of positive and negative particles with high electrical conductivity where the component particles are close enough to interact intensively.
2. The energy unit of 1 eV is equivalent to 1.6×10^{19} watts per second (Wsec).

3. Initially ITER, but more recently revamped as "The Way," from Latin *iter*: www.iter.org, and "Why Fusion Power" at www.iter.org/a/index_nav.
4. Isotopes of any type of atom contain the same number of protons but differ in the number of neutrons. Three types of hydrogen isotopes are known, ordinary hydrogen (just one proton), deuterium (one proton and one neutron), and tritium (one proton and two neutrons). These heavy isotopes are chemically identical to the ordinary hydrogen atom.
5. For an historic overview, see www.mbe.doe.gov/me70/manhattan/cp-1 critical.htm, nuclearweaponarchive.
6. H. Frauenfelder and E.M. Henley, *Nuclear and Particle Physics*, Benjamin, Reading, Mass., 1974.
7. R. Dorf, *Energy, Resources and Policy*, Addison-Wesley Publishing Co., Reading, Mass., 1978.
8. P.A. Tipler, *Physics*, Worth Publishers, New York, 1991.
9. See, for example, http://news.bbc.co.uk/2/hi/science/nature/6165932.stm and http://news.bbc.co.uk/2/hi/science/nature/5012638.stm.
10. The 2,000 MW difference comes about in the transition from thermal to electricity.
11. www.jet.efda.org/pages/content/news/2005/yop/oct05.html; and the original 1997 press release, www.jet.efda.org/documents/jetpressrelease/press-release-1997.html.
12. Presentation by B.D. Wirth, www.nuc.berkeley.edu/courses/classes/NE39/Wirth-FusionMaterials_lecture2.pdf, and S.J. Zinkle (2004) p. 47, http://fire.pppl.gov/aps_dpp04_zinkle.pdf.
13. Ibid.
14. The construction and operation timeline details (last accessed December 26, 2007) can be found at www.iter.org/a/index_nav_1.htm; www.iter.org/a/n1/downloads/construction_schedule.pdf.
15. "Elastic" scattering of neutrons is like billiards, where in collision the original balls do not get destroyed; there is just momentum exchange. "Inelastic" refers to a situation where all or some collision "partners" get destroyed.
16. For an interesting description of some of these difficulties encountered in small-scale experiments, see J.L. Anderson, "Tritium Systems: Issues and Answers," *Journal of Fusion Energy*, vol. 4, nos 2/3, 1985, www.springerlink.com/content/m344456872521544/.
17. This is relatively small compared to today's standard 3,000 MW (therm) fission reactors which achieve up to 95 percent steady-state operation.
18. See, for example, M. Abdou, "Notes for Informal Discussion with Senior Fusion Leaders in Japan (JAERI and Japanese Universities)," March 24, 2003, www.fusion.ucla.edu/abdou/abdou%20presentations/2003/Japan Notes%20for%20Informal%20Discussion%20FINAL%203-17-032.ppt.
19. Ibid.
20. www.fusion.ucla.edu/abdou/.
21. M.E. Sawan and M.A. Abdou, "Physics and Technology Conditions for Attaining Tritium Self-Sufficiency for the DT Fuel Cycle," *Fusion Engineering and Design*, vol. 81, nos 8–14 (2006), pp. 1131–44.
22. The fractional tritium burn-up during the short MW pulses in JET was roughly 0.0001 percent.

23. Sawan and Abdou, "Physics and Technology Conditions for Attaining Tritium Self-Sufficiency for the DT Fuel Cycle."
24. Ibid.
25. See, for example, S. Balibar, Y. Pomeau, and J. Treiner, "La France et l'énergie des étoiles," Point de vue, *Le Monde*, October 24, 2004; W.E. Parkins, "Fusion Power: Will it Ever Come?" *Science*, vol. 311, no. 5766 (March 10, 2006), p. 1380, and http://fire.pppl.gov/fusion_science_parkins_031006.pdf
26. M. Abdou, "Briefing to DOE Office of Science on Fusion Chamber Technology," Washington, June 3, 2003, www.fusion.ucla.edu/abdou/abdou%20presentations/2003/orbach%20pres%20(6-1-03)%20Final1.ppt.

19
Geothermal

Sheila Newman

Both the sun and the earth condensed from dispersed star matter in a process that generated heat and left both very hot, the sun more so. The sun would have cooled down very long ago but for the energy generated by hydrogen fusion. Like the sun, the earth started hot but would also have cooled down long ago if by itself spinning in cold space. Radiant heat from the sun reduced the earth's loss of heat to space while the heat generated by the radioactivity of uranium, thorium, and, particularly, potassium in the earth has also maintained the internal temperature.

There is considerable conjecture as to the relative contributions of primal heat associated with planetary formation, radioactivity, and solar radiation to the balance, but the one thing we know for sure is that the earth's core is very much hotter than its crust. There is a negative temperature gradient from the core inside to the crust. It follows, then, that the deeper you drill, the greater the temperature differential to drive a geothermal project, bearing in mind that the thermal conductivity of different rock types varies and alters the heat outflow picture according to geology.

A fluid (or even air) is heated below the surface, brought up, and used to generate electrical power. Geothermal energy supplies power in many countries around the world, particularly in areas where it is obviously close to the surface. Iceland, notably, has switched its dependency almost 100 percent from fossil fuels to geothermal electricity over the past 50 years. Some of this energy is stored in hydrogen and used for transport in this process where electricity is comparatively cheap and does not require fossil fuel, but much of Iceland's transport energy is still fossil fuel based.[1]

It should not be a surprise that the entire world resource base of geothermal energy has been calculated in government surveys to be larger than the resource bases of coal, oil, gas, and uranium combined. In fact, potentially accessible geothermal resources have been calculated in an MIT study to be sufficient to supply current planetary needs for many thousands of years. The figures cited have been 13,000 zettajoules (ZJ)[2] of which around 200 ZJ would be recoverable with possibilities of increasing to 2,000 ZJ.[3]

The need to drill deeper and deeper for oil means that oil companies now drill to depths of 10 km almost routinely. Geothermal to this depth requires not-too-dissimilar technology. For comparable initial financial and energy investment much vaster and, moreover, renewable, thermal energy resources will theoretically become accessible.[4] In the meantime, why not use the wells we already have?

> Using coproduced hot water, available in large quantities at temperatures up to 100°C or more from existing oil and gas operations, it is possible to generate up to 11,000 MWe of new generating capacity with standard binary-cycle technology, and increase hydrocarbon production by partially offsetting parasitic losses consumed during production.[5]

> It has been suggested recently that there is an enormous untapped hydrothermal energy resource associated with coproduced hot waters from oil and gas operations (McKenna and Blackwell, 2005; McKenna et al., 2005). Those authors estimated that the resource potential could range from about 985 to 5,300 MWe (depending on the water temperature), using the fluids currently being produced in seven Gulf Coast states.[6]

Although these energy sources lose their heat over time, if left to lie "fallow" for around ten to fifteen years, the heat is renewed by conduction from deeper in the earth and from radioactive heat in surrounding rocks. The idea is to work a well for about five years at a time, and not to allow stored heat to drop more than about 20 degrees Celsius. Wells would generate about 50–100 MW each, but there would be several in each field, spelling each other.[7]

HOT ROCKS GEOTHERMAL ENERGY

This is a new variation on drilling for geothermal energy. "Hot rocks" projects, such as one in South Australia,[8] involve heating water passed through slightly radioactive granite, which isn't the same thing as simply accessing heat by descending far enough into the earth's crust. The South Australian "Geodynamics" project is planned for a depth of just under 5 km.

The natural radioactive elements are concentrated more in "acid" granite than other rocks. This is why granite, capped by less thermal conducting sediments, is a target for hot rock ventures in South Australia.[9]

Another "hot rocks" project exists in Strasbourg, France. A project in Switzerland, located on a fault which destroyed the medieval town of Basle in the fourteenth century, has been seriously delayed or halted by being linked to earth tremors.[10] The Australian site has the advantage of being located in a flat, geotectonically stable desert, whereas the European sites are relatively close to dense populations. None of these projects was yet producing commercially at the time of writing.

Geothermal energy appears to carry much fewer traditional environmental problems than other energy sources. Plants are smaller and localized. There isn't the same need to refine and process fuel before it becomes usable. Potential environmental and social impacts include highly poisonous hydrogen sulphide gas (H_2S). Carbon dioxide occurs in the geothermal stream at the rate of about 4 percent of that released by fossil fuel plants, but release may be avoided in a closed cycle. The dissolved mineral content is higher than in cold groundwater aquifers and needs to be segregated. Geothermal plants are local to their source and tend to be much smaller than other power plants.[11]

The greatest conventional threat it would seem is the impact from human activities and infrastructure for cities likely to rise around geothermal sites. The massive scarring of the earth's surface which occurs in coal mining and oil drilling and refining, and the pollution entailed in uranium mining, are absent.

That said, if deep wells were drilled for geothermal energy at similar rates for mining of other forms of energy, the seismic impact would almost certainly become relevant, perhaps opening up a hitherto undreamed of physical dimension of complications. (One cannot help but think of the Sorcerer's Apprentice.)

The risk of increasing or initiating seismic activity sufficient to create earthquakes is not a negligible risk. Professor Tester of MIT suggests that by drawing energy out of the earth's crust it may, however, be more likely that seismic activity will decrease rather than increase near geothermal projects. Australia's geological stability and relatively unpopulated hot desert interior seem intuitively more suitable than Basle with its violent seismic past, but seismically active countries are more likely to think geothermal. Cases in point are Iceland, described as "almost a living earthquake" by Prof. Tester, and California, which has the world's biggest wastewater-to-electricity geothermal project, pumps sewerage via 40 km of pipes into the Guysers geothermal field. The twice recycled by-product is to be redistributed to local irrigators.[12]

New energy supply systems based on new energy sources have the potential to change geopolitical destinies and alliances. Kenya, Ethiopia, Djibouti, Uganda, Tanzania, and Eritrea, situated along the African Rift Valley, which holds huge geothermic promise, are currently seeking funding from the World Bank to develop this. Estimates have the potential power at up to 7,000 megawatts (MW) – the equivalent of 7,000 conventional power plants.[13] Admittedly it stretches the imagination to surmise that the powerful nations of the world will not steal this too, but the local nature of geothermal electricity production might be of more benefit to Africans.

Geothermal energy is on a planetary scale right under our noses and only its most obvious manifestations near the surface have yet been commercially accessed. In theory it dwarfs all other energy sources to date, but deep geothermal wells are still experimental. The greatest problem is to gain control over the size and shape of the geothermal reservoir artificially created by fissures and to effectively recover the water sent down the well.[14] That is no small problem.

NOTES

1. Karl Gawell and Griffin Greenberg, "2007 Interim Report Update on World Geothermal Development May 1, 2007," GeoEnergyOrg, http://tinyurl.com/2e38u4, accessed October 19, 2007.
2. ZJ stands for zettajoule, which is 10^{21} joules.
3. "The Future of Geothermal Energy," Idaho National Laboratory, http://geothermal.inel.gov/publications/future_of_geothermal_energy.pdf. See also an excellent online lecture from MIT Prof. J. Tester at http://techtv.mit.edu/file/240/.
4. For an overview of recently begun projects, see Gawell and Greenberg "2007 Interim Report Update on World Geothermal Development May 1, 2007."
5. Massachusetts Institute of Technology, *The Future of Geothermal Energy*, MIT Press, Cambridge, Mass., 2006 (ISBN 0–615–13438–6), http://geothermal.inel.gov/publications/future_of_geothermal_energy.pdf, p. 20.
6. Ibid., p. 241, accessed December 26, 2007.
7. Prof. J. Tester, MIT, "Geothermal Lecture," http://techtv.mit.edu, accessed November 17, 2007.
8. "Energy from Hot Rocks – An Example of the Hurdles for Innovators," Interview with Prame Chopra by Robyn Williams, *The Science Show*, May 19, 2007, www.abc.net.au/rn/scienceshow/stories/2007/1927118.htm.

9. Ibid.
10. E. Engeler and A.G. Higgins, "Energy Search Goes Underground," Associated Press Writers, August 4, 2007.
11. Geothermal Education Office (GEO), "Geothermal Energy Facts," http://geothermal.marin.org/geoenergy.html#future, accessed December 26, 2007.
12. http://tinyurl.com/yrowub, accessed October 19, 2007.
13. The African Rift Valley Geothermal Development Facility (ARGeo) project. Gawell and Greenberg, "2007 Interim Report Update on World Geothermal Development May 1, 2007."
14. *The Future of Geothermal Energy*, chapter 4.

Part IV: After Oil

> Complex societies, it must be emphasized again, are recent in human history. Collapse then is not a fall to some primordial chaos, but a return to the normal human condition of lower complexity.[1]

Much that is written about oil depletion imagines the future almost entirely from a US viewpoint. Even for North America, a US focus is almost certainly too much of a generality. How long will the concept of a nation composed of 52 states endure oil depletion, before it becomes 52 countries and then many more bio-regions? Steven King's post-apocalyptic horror/science-fiction novel *The Stand*, originally published in 1978, might still set the scene.

Will history record a time, however brief, when the Western European states truly merged into a European Union, or will fossil fuel depletion shrink them back into their eighteenth-century polities and bio-regions too quickly?

Looking away from the US or a United Europe, this part invites the reader to approach a post-fossil fuel future from the slightly parallax perspectives of France, Australia, North Korea, and Japan.

The treatment of three of these countries involves projections based on trends. The third country, North Korea, involves a history rather than a projection. North Korea, in fact, lost access to cheap oil with the fall of the Soviet Union. So did Cuba. Although not specifically treated here, Cuba has arguably become an example of community triumphing over petroleum scarcity by "relocalization" and land redistribution. In contrast, North Korea is a tragic lesson about the isolation of a monolithic state by its enemies and rigid leaders in a sea of plenty. The chapter on Japan explores potential similarities and differences between North Korea and Japan.

These chapters on North Korea and Japan, by Antony Boys, make exotic statistics accessible and were rewritten from much longer works specifically for this volume.

As the writer of the chapter comparing and contrasting the post-peak potentials of France and Australia, I am very pleased to be able to publish the work of a fellow spirit. Indeed, I can imagine a whole volume devoted to "future scenarios" for different polities and regions and the political outcomes of their following different technological routes.

I feel that this is an evolving field which may bring together people from all over the world to make both interesting and important contributions to our future.

In "The Simpler Way," Ted Trainer echoes and develops most of the ideas in this book to a not-unpleasant conclusion: that simple is not stupid, and is probably better.

The final chapter, about thermodynamics and nature, follows on from the ecosystem approach of the earlier "Peak Soil" and "Notes on Terra Preta" chapters (Part III), and reminds us that The Final Energy Crisis isn't just about us humans.

NOTE

1. J. Tainter, *The Collapse of Complex Societies*, Cambridge University Press, Cambridge, 1988; 12th reprint, 2005, p. 198.

20
France and Australia After Oil

Sheila Newman

INTRODUCTION

This chapter explores the future of human society in France and Australia after fossil fuels, assuming a "petroleum interval" lasting from about 1850 to 2035. The method is to try to reconstruct the situation in each country before fossil fuel, then to extrapolate according to trends, physical and social.

Given that there are different ways of "powering down" from peak oil, this chapter is as much political as geophysical.

Biophysically very different, Australia and France have historical similarities and differences stemming back to 1788 and 1789 respectively, when new European political regimes were installed; one colonial, the other revolutionary. A British-based system of land tenure and expansive politics shaped an Australia vulnerable to land speculation and privatization. In France a series of Republics commenced with the 1789 Revolution, culminating in the current (Roman-based) Napoleonic system of land tenure and inheritance, discouraging land fragmentation, with taxes penalizing speculation, housing a civil right, and a multi-party system. Both polities were shaped by the rise of coal. Both became manufacturers and nearly self-sufficient in the nineteenth and twentieth centuries. With the rise of petroleum, both aimed for population growth and infrastructure expansion. The 1973 oil shock, however, saw France rein back population growth (both natural increase and worker immigration) and related drawdown of materials and energy, whilst Australia (after initial attempts to stabilize population) fell victim to a more predatory capitalism,[1] accumulating national and personal debts in a Thatcherite economic climate which sacrificed community, manufacturing, and political independence for commodity dependency and subserviance to US-Anglo hegemony, in a corporate reworking of colonialism.

On December 4, 2003, Australia's population was estimated at 20 million and projected to reach about 30 million by 2050. Slightly less than 50 percent of this growth rate resulted from net overseas immigration. Updating this edition on November 5, 2007, I find that Australia's population has ballooned by more than one-twentieth of itself (or 5.66 percent), to 21,131,216, and is projected to reach 34 million by 2050.[2] In fact, with that growth rate of 1.5 percent per annum, it is on course to double within less than 50 years. Annual immigration was responsible for more than half this growth, even though the birth rate had increased in a context of misleading pronatalist propaganda.[3]

By contrast, France's population at 61,538,000 on January 1, 2007, had grown by 2.56 percent at 0.64 percent per annum – up from 0.39 percent per year prior to 2003. A large part of this growth was due to natural increase, particularly from older mothers. After 2050, with the demise of most baby-boomers, France's population is still set to decline.[4]

FRANCE

A quick way to gauge carrying capacity is to look at the size and vigor of the animals and plants in an area. Europe has the capacity to sustain many humans plus 27 species of mammalian carnivores, including two kinds of bears (the biggest, most energy-intensive mammal) in an area not much bigger than Australia. Europe and France were blessed with a mostly agriculture-friendly climate and very rich soils, due to recent glaciations grinding and renewing the earth as they withdrew only 8,000 years ago. Despite these natural advantages, modern agricultural equipment, irrigation, single-cropping, and extreme reliance on mineral fertilizers has radically increased erosion and soil degradation[5] all over Europe, compounded by increasing rainfall in regions north of about 45° N latitude.

Isolation or its lack is a crucial factor to be considered in projecting the future of complex societies under changed circumstances.[6] After the collapse of Rome, no European society completely crashed, probably because there were many independently resourced neighbors and competitors able to absorb parts of failing polities or to recolonize their remains.

Before the introduction of coal as a widespread substitute for wood, around the time of the 1789 Revolution, France was a self-sufficient agricultural country, with a high proportion of cereal crops. In conceptual terms, a return to later nineteenth-century land use and farming intensity, if not techniques, and to early- to mid-nineteenth-century population numbers, may not be as difficult as it might look, because we need only to go back to the well-documented past.

Population

From medieval times until the middle of the eighteenth century, France's population, the largest in Europe, oscillated around 18–20 million. Between 1750 and 1815 it reached nearly 30 million.[7] Expansion through warfare also increased France's population by altering its borders: in 1850 Nice and Swiss Savoy were added, with nearly 1 million more people and their territory. Increases in land by warfare, drainage, irrigation, and other works from the middle of the nineteenth century were, however, more than counterbalanced by loss of better land to urbanization, soil degradation, and to pollution from industry and intensive agriculture.

The French industrial revolution did not begin until around 1880. World War I, the Great Depression, and World War II further delayed this development. Compared to Germany, the UK, and the US, France had only modest coal deposits. For this reason and others, French population growth was moderate relative to growth in Germany, the UK, and the US. Over the period of "nation-building" in Europe in 1780–1860, in which railroads, post, and telegraph systems played a major part, France increased its carrying capacity through wealth and commerce arising from foreign possessions, especially in the cities. French colonies, however, did not experience the booms of British ones, probably because they did not have Britain's population "surpluses." Otherwise they might have kept West Australia.[8]

The relatively secure post-Revolution French peasantry, with strong cottage industries, had little reason to settle France's colonies, work in factories, or serve in the army. Their land tenure security meant that France lacked Britain's hoards of dispossessed laborers, a landless caste maintained from medieval times. French industry and the army therefore resorted to luring immigrants. So, largely due to immigration, France's population grew from 29.4 million to 35 million between 1815 and 1845, and then from 35.6 million to 38.4 million between 1850 and 1869.

In the late nineteenth and early twentieth centuries, the primarily rural population of France was often little integrated into the growing urban and industrial-based cash economy. Peasants still used candles and mineral- and animal-oil lamps for light. Horses, donkeys, oxen, and even cows and dogs were still widely used for road transport and hauling. The dominant industrial energy sources remained the water wheel, windmills, and tide mills. Rivers provided the most energy-efficient form of transport with networks connecting to those of other European countries. Bulk transport, wherever possible, was by boat and barge.

Today, large-scale hydropower produces about 70 terawatt hours per year (TWh/yr), and small-scale (below 150 kilowatts (kW) and run-of-river hydroelectric installations produce about 4.5 TWh/yr. France also has one of the world's few operational, large tidal electric power stations (Rance River, Brittany). In the late nineteenth century coal was increasingly used, but wood remains a major source of commercial and non-commercial energy, still providing around 40 percent of rural space-heating fuel requirements. Furthermore, French law requires all apartment buildings to retain chimneys for wood burning in case of interruption of other fuel supplies.

Primary productivity gains are dependent on the fossil fuel economy. While the French may claim the status of "the EU's breadbasket," this ignores two key factors. The first is the pollution and depletion of water tables – most spectacularly for pollution in Brittany, and most intensely for depletion in the entire south and southwest of the country. Second, although French agriculture is among the most productive in the world, it is heavily dependent on fossil fuels and petroleum products – both directly for machines, and indirectly for fertilizers, insecticides, animal medication, and other inputs necessary for intensive production. Once these are stripped away, food self-sufficiency for around 25 million people – less than 50 percent of France's twenty-first century population – becomes an optimistic but perhaps attainable target.

When I wrote the first version of this chapter in 2003, trends and social structures in France still promoted consolidation against population growth. The Napoleonic Code of 1804, widely imitated throughout Europe, had reinforced and introduced inheritance laws which continued to counter land fragmentation and indirectly mitigated against profiting from population growth and land speculation. Writing in December 2007, I note with apprehension that new French President Nicolas Sarkozy has introduced many changes to these wise laws which will weaken France's ability to resist the population growth and land speculation lobbies that currently afflict the English-speaking countries. The French are naive about the impact of these changes, but corporations, lawyers, and notary beneficiaries are not.[9]

Future

Nuclear electricity was established by de Gaulle after World War II, consolidated during the first oil shock, and remains the first choice of current policy makers. It will last as long as uranium supplies – probably only until 2040 rather than to 2200, depending on demand and supply.[10] France, like other nuclear powered countries, hopes to make a transition to fast breeder reactors (FBRs) after uranium. It is also caught up in the great fusion experiment (ITER).[11] There is serious but limited experimentation in "hot rocks" geothermal.[12] Will these hopes, plus trends towards privatization emerging under President Sarkozy in 2008, detract France from increasing its efforts to seriously supplement nuclear with flow energies, in the very likely case that FBRs, geothermal, and (especially) fusion don't work out? So far France has retained a lot of dirigiste character and, as Colin Campbell suggests in his chapter "Caspian Chimera" (referring to Soviet oil exploration),[13] state involvement in financing and planning new technology and energy choices is potentially far more efficient than *le capitalisme sauvage.*

Much of France's pre-fossil-energy infrastructure either exists or could be reconstructed, or even improved upon. This is particularly true of the canal system and woodlands. France's woodlands and forests have been maintained, and have even increased on earlier times. Efficiently used, in combined heat-and-power facilities, wood and other biomass energy resources can easily provide reasonable heating, and sensible electrical energy needs of an equilibrium or sustainable population of around 20–25 million people. The capacity of the managed forests of France to support the return of a functioning biodiversity, from which people could supplement food sources (small game) and obtain fuel, would probably require changes in the vegetation species mix for best adaptation to regional climates and soils. In the future, there may also be potential for wind-powered transport, both along canals and rivers, as well as roads.

The ability to cover wide areas with the fast transport, and now communications, necessary to bind disparate communities and cultures into what are called modern nations is the hallmark of fossil fuel and will of course diminish and change without it. But it might be possible for France to maintain electricity supplies for a certain level of high- or medium-speed rail transport[14] through a mixture of renewable energy sources – that is, water, wind, wood, and tidal energy.

On balance, a population of around 20–25 million might be potentially sustainable in the post-fossil-fuel and post-uranium era. If France applied soil conservation policies and practices, recycled soil nutrients, and recovered biological diversity, pre-nineteenth-century productivity might be restorable in some regions. This could then support the higher end of the very rough, sustainable population estimate.

The twentieth-century infrastructure overlay to a target population (20–25 million), which is the same as that at the beginning of the nineteenth century, offers some potential for restructuring, adaptation, and continued use. This notably includes rail transport. At present, some all-electric, but mostly diesel-fueled onboard electric generator-powered trains link major points of the country and neighboring countries.

The infrastructure is also still largely in place for canal traffic at a few percent of current road transport capacity. Paris has organized a system of free public bicycle travel and government and employer subsidies for rail.[15] Beyond some unsustainably large and energy-intensive conglomerations, the spatial organization of human settlements in France remains pedestrian-friendly, with space between settlements easily covered by boat, horseback, bicycle, or on foot. Other than canal and maritime transport, rail transport is the most energy-efficient. Building stock, insulation, and design improvements could be applied to selected building groups in efficiently and rationally located settlement centers *outside* today's major urban areas, such as Paris – Ile-de-France, Marseilles, Lyon, Lille, and Bordeaux, enabling a stabilized and decentralized population to live in an increasingly sustainable way. Many entire villages contain historic but viable building stock, adaptable to much lower energy operation. Indeed, since the first oil shock, French policy has encouraged the re-equipping, renovation, and adaptation of old building stock, with the result that there has been a consolidation of decentralization trends.[16] French national and regional parks have made a feature of preserving the viability of customs and industries in villages which might otherwise have fallen into ruins.[17] France famously protects its agriculturalists and has not become dependent on imported food.

It should thus be possible to return to the use of horses and other beasts for transport. If all else fails, it might even be possible to use animals, wind power, and water power to draw light trains along rail lines, as in the nineteenth century. The critical question will remain the *maintenance of energy-intensive infrastructures*, which at present are entirely, or substantially, dependent on fossil fuels.

The position and logistics of energy-producing resources, including land for food production, and how equitably these and their products are distributed will shape the infrastructure design and population distribution (as well as the numbers) and the "social contracts" that new regional or national entities will develop in the period from about 2035.

AUSTRALIA

Before British colonization in 1788, the peoples of *Terra Australis* managed to conserve an almost exclusively hunter-gatherer nomadic lifestyle. Art,[18] but no written history, has been found, and reconstruction of their impact relies on anthropological, archeological, and ecological studies. "Australia" was transplanted and adapted from a British society which was on the cusp of industrialization. Pre-1788, Australia's Aboriginal population *averaged* continent-wide fewer than one person per 8.5 square kilometers – possibly as few as one person per 51 square kilometers.[19] Numerous clans inhabited the continent at different population densities, reflecting regional rainfall, soils, and climate.[20] Also patterned by climate and soils, the fossil-fuel-era population distribution is similar, but much *denser*.

Early attempts to establish agriculture failed, with some unintensive exceptions recently uncovered.[21] The British managed to gain an agricultural foothold using "white" slaves in the form of convicts drawn mostly from the ragged army of their dispossessed. Their number was later supplemented by indentured labor, displaced Aboriginals, and, until Federation, "blackbirding" – the practice of kidnapping Pacific Islanders and bringing them to work in Australia, principally for the Colonial Sugar Refinery Company (CSR).

There is thus no history or tradition of an established pre-fossil-fuel agricultural society. The gold rushes of the 1850s attracted capital, finance and economic migrants, resulting in a rapidly morphing population and economy and the formation of a working class. This class made a national wage-fixing pact with capital at Federation in 1904 and also obtained the agreement of the CSR to outlaw blackbirding,[22] and the importation of other "non-white" labor, widely perceived as synonymous with slaving.[23]

The economy intensified after World War II, but much land was cleared and divided up for development by land speculators from the time of the gold rushes of the mid-nineteenth and early twentieth centuries. When the gold ran out, there was a massive depression, which probably assisted the formation of the above industrial laws.

After World War II, business promoted a fear of population implosion among politicians[24] and a policy for mass immigration came in. High immigration, combined with the unforeseen baby boom that accompanied the petroleum era, made the newly privatized housing industry very powerful and consolidated an economic addiction to population growth. Although the "white Australia" policy was dismantled, wages and conditions legislation under the 1904 constitution protected workers and made it unprofitable to import labor simply to undercut wages. However, in 2006–07, the conservative government found a way around this – "Work Choices."[25] At the same time net immigration was encouraged to increase from an average of around 75–80,000 per annum to upwards of 160,000 per annum,[26] at the behest of the development, housing, mining, and financial lobbies. All this took place in the context of a huge increase in mining and construction, including massive engineering projects in most states which have drawn angry but useless protests from Australians. These circumstances underpin Australia's demographic and material overshoot.

Biophysical

Australia has the capacity to support a very small but very varied biomass, compared to Europe.[27] Reasons for Australia's lesser carrying capacity are its unreliable rainfall – "drought and flooding rains"[28] – hostile climates, and impoverished soils, which, unlike Europe's, have not benefited from recent glaciations or volcanic activity.

Australia's noteworthy flatness discourages rainfall and allows rivers to meander and evaporate, salting up soil, but gathering and depositing little silt. Non-nomadic Australian mammals are small, with slow metabolisms. The largest mammals are the nomadic kangaroos. Like cockatoos, parrots, and the long-legged flightless emu, the kangaroo can travel quickly away from drought to rain, to find better foraging. Many of the largest fauna are reptiles – monitors, snakes, crocodiles – feeding infrequently, using the sun to help maintain body temperature.[29] Flora and fauna have adapted to the precarious, infertile soils, and mostly arid climate. For instance, Australia has the world's largest variety of carnivorous plants, which supplement nitrogen-poor soils with nitrogen-rich insect protein, and *Nuytsia floribunda*, a tree-sized mistletoe, draws nutrients directly from the roots of trees.[30] The "scleromorphous" structure of much indigenous vegetation is an adaptation to water shortage.

Most of Australia is hot desert. Total land stock is 770 million hectares, but less than 30 million hectares (below 4 percent), is of good or very good quality for cropping.[31]

Some 75 percent or 570 million hectares is rangeland. About 406 million hectares are used for grazing, with stock density as low as one beast per 100 square kilometers. Frequent droughts lasting several seasons often result in massive stock deaths. The majority of Aboriginal people moved with their nomadic food sources in response to the erratic climate and changing conditions. In some more fertile, less arid parts, especially the southeast, archeological remains of stone habitations and signs of eel races were found recently, indicating more settled conditions at some stage.[32]

European impact on "productivity"

Fossil fuel based, mechanized, extractive economies are a bit like glacial and volcanic action, bringing new components from deeper layers to the earth's surface.

Because of this, since 1788, continental biomass, or "net primary productivity," has probably increased by around 5–7 percent.[33] This increase has been restricted mostly to about one-quarter of the country's surface area, mainly in the southeast. Based on this 5–7 percent range and the location of intensive production in Australia, it seems reasonable to suppose that net primary productivity has been concentrated and therefore increased in this more fertile quarter of the continent by about 20 percent, and that productivity in the most fertile 4 percent would have increased by over 100 percent, mostly due to fossil fuel based synthetic fertilizers, irrigation pumps, and fuel-burning engines for machinery and transport.[34]

How many people in what kind of future economy?

There is an old joke which goes that Australia could have a population as large as Bangladesh's – for a few days. Prior to European settlement, the population of the Aborigines, estimated at between 300,000 and 900,000,[35] survived tens of thousands of years. In 219 years, the population base founded by Captain Cook, with its technologies and imported animals and crops, is in overshoot in a number of areas and will have great difficulty maintaining itself, let alone future generations, in the absence of fossil fuel. Climate change threatens additional disruptions and desertification.

Creeping dry-land salinity affects 2.5 million hectares. Some 17 million hectares of the 30 million total of good land are likely, on current trends, to be destroyed by salinity by 2050, leaving 13 million "good," and trending to a halving of agricultural productivity.[36] More than 24 million hectares of soil are considered acidic. Much of this is natural, but agricultural management technologies are causing the soil acidification process to accelerate.[37]

National policy is to export the rest of Australian coal and gas to bolster rapid economic growth without serious regard for environmental and social costs. At this rate Australia's remaining fossil fuel resources, along with those in the rest of the world, will rapidly deplete.

Short term to 2050

It seems that there are two choices available in the short term: the "Cuba route" and nuclear power.

The Cuba route: Anthony Boys's chapter, "North Korea: The Limits of Fossil Energy-Based Agricultural Systems," in this volume discusses how badly North Korea fared when cheap soviet oil was no longer available at the end of the Cold War. Another story, better known, is how Cuba dealt with this situation.[38]

The biggest problem for Cuba was and remains transport. Bicycles can only do so much. Servicing a population and infrastructure shaped by oil but without that oil is fraught with difficulty. Cuba had relied on importing food, agricultural fertilizers, and petroleum for road transport. Relocalization of institutions as well as food production was the main solution. People in the cities were assisted to produce their own food and state farms were turned into small leaseholds and cooperatives. Permaculture techniques and draft animals replaced fossil fuel-based fertilizers and petroleum for farm machines and transport. Cuba is now self-sufficient and Cubans purchase 80 percent of their food at local markets. Throughout and still, the state remained a vigilant welfare provider, attending particularly to health and education, with an unusually high ratio of doctors per capita.

The key to the Cuban solution was land redistribution, which was possible within Cuba's system. This alternative is not going to be easily available in Australia because of the land use planning and housing system, which makes relocalization very difficult. That is a pity because many see this alternative as a desirable way of profiting from adversity by reducing centralization and aggregation of land, energy resources and utilities, and strengthening democracy, and independent and collective economic and social participation.

The nuclear route is the opposite way. It implies total electrification, synthesizing, using complex technology, the pattern of settlement and transport which arose from exploitation of the natural endowment of petroleum. It requires massive investment, and in Australia's case the investment sought is likely to be private. This implies ongoing loss by citizens of control over the country's energy systems and all that flows from that – the future of work, the state of the environment, natural amenity. It implies the reshaping of Australian society by corporations with profit alone in mind, for the benefit of a small, dominant, asset-rich class, with the electorate a mere captive market.

In Cuba the energy supplies and materials which permitted centralized power were suddenly withdrawn and Cuba lacks the mineral resources to attract capital intensive investment. This situation surely facilitated local reorganization. In uranium-rich Australia, an opposite structuring is more likely. Because government policy favors commercial investment over public investment, nuclear facilities would tend to consolidate corporate power and its relationship with national and state government. Australia's Anglophone mainstream media derives largely from syndicated US sources and these cultural mediators not only do not support communism or socialism, but they do not support dirigisme or relocalization either.

In the space of four years, long-established Australian government anti-nuclear power policy has reversed in spite of public disapproval. Nuclear is seen as an investment opportunity which will provide employment, international importance, and new industries. Rationales offered to the public are the imperative to provide power for projected (politically engineered) population growth and the desirability of offsetting greenhouse gas contributions from coal-fired electricity and coal exports. (Australia is the world's largest coal exporter.) Australian planning is dominated by what the corporate sector wants. There are many indications that public sector scientists are expected to support private, corporate research and development rather than leading with public research responding to public need, which might result in moderation rather than accelerated consumption.[39]

According to the ABARE,[40] in the financial year 2005–06 Australia's total primary energy consumption was 5,640.7 petajoules (PJ),[41] growing at an average of 2.2 percent or about 120 PJ per annum over the past 20 years. No one is seriously suggesting replacing all this with nuclear. The suggested figure is more like 25 power plants in the next 15 years supplying one-third of Australia's (doubled) electricity demand.[42] But many people believe that nuclear will actually take care of the problem of fossil fuel depletion, permitting business as usual.

To do so would theoretically take 225.6 1,000-megawatt-electric (MWe) nuclear power plants each supplying 25 PJ,[43] plus an average of 4.8 new nuclear plants per year to supply annual growth in consumption. Each plant's initial cost would be around US$2 billion, in an economy currently valued by the World Bank at US$768 billion, but heavily indebted and privatized.

Of the 5,640.7 PJ primary energy, around 55 percent comes from petroleum (35 percent) and gas (19 percent).[44] To rely on nuclear to replace this by powering an electric car or hydrogen road transport system is a further absurdity. In addition to the 225.6 imagined nuclear power plants, more would be required to supply electricity to plug-in electric cars and to split water into hydrogen to carry more electricity to fuel the creation of new infrastructure to manufacture and service hydrogen-carrying electric vehicles, requiring more electricity (obviously).

Each 1,000-MWe nuclear power plant would itself require around 1.3 PJ per annum or 52 PJ over a 40-year lifetime according to the Australian Uranium Association.[45]

The government is counting on coal and gas to fill the initial gaps in petroleum supply, but coal is being exported so fast and the world is using it up so quickly with the gas peak already looming, that nuclear will presumably be called upon to supplement coal, gas and oil as well. Perhaps the government is desperately trying to attract nuclear development money by enlarging its population base and intensifying its other extractive industries, in a plan where uranium mining is, of course, a keystone. None of the above calculations factor in the amount of new energy required to build the cities around the plants – to supply and service the infrastructure required for the projected population growth, which is used to lure investment.[46]

In a totally electrified economy, attempting to preserve something like the current consumer goods-based economy, the most likely option is to produce hydrogen electrically and also use it as an electricity carrier, to fuel electric cars and trucks. This implies a vast increase in the need for electricity and a complete renovation of road transport support infrastructure as well as vehicle design and manufacture. The corporations will demand profits from any new ventures. How are the Australian people going to pay for this? What are they going to need to give up for this? In Australia's "trickle-down economy," who is going to be left out? At the very peak of oil production, public priorities have shifted so far in favor of private enterprise that the Australian government is not capable of providing affordable housing for its constituents.

In the meantime the VAMPIRE index (Vulnerability Assessment for Mortgages, Petrol, Interest Rate Expenditure) is rising.[47] The people at the high end of the VAMPIRE index are those locked into bloating mortgages in remote suburbs with little or no public transport, captive to rising gasoline prices for cars without which they cannot get to work, get their children to school, or get to the shops to buy food.

Such hostages to sprawl are most in need of the Cuban route, but Australia's systemic inability to plan and adapt infrastructures, industry, and national resource use to radically changing conditions is perhaps its biggest problem.

Planning departments and tribunals palliate a bewildered electorate with superficial choice between styles, but, whilst allowing endless sprawl on the city edges, they continue to ram through intensified developments in the cities, and purposefully to overshoot water resources and transport infrastructure in response to the growth lobby's insatiable demand for big projects at public expense. These take the form of desalination plants, sewerage recycling plants, complex road tunnels and expressways, and, biggest of all, geological restructuring of seaports to accommodate giant container-ships for anticipated quadrupling in trade.[48] With chilling disregard for the consequences, Australian lobby groups of planners, developers, builders, and financiers have successfully campaigned for government-led policies to increase population growth and energy demand.

One of the greatest problems is that soil and water damage are caused by industrial agriculture, but because industrial agriculture nets huge profits in the short term, and depends on huge markets (big populations), capitalist governments are not at all motivated to stand up to corporations who want laws which tend to make it harder for people to act locally and find autonomy.[49] Any technology offering a lot of electricity, like nuclear power, would permit continued over-use of nitrogen-based fertilizers for monocultures, which so suit the corporate economic and social structure.

Biofuels are ecologically and economically very expensive (see "Peak Soil" by Alice Friedemann, this volume) but very attractive to big business, in part because of the initial look they have of offsetting carbon gas contributions.[50] It is more likely that they could be used moderately, in a local setting, because their cost and value would be obvious to any permaculturalist and they would therefore be used very judiciously (perhaps for small farm machines and transport).

Solar collectors and wind on a systemic scale demand intensive funding. Even to support the development of the technology requires intensive funding. Whilst uranium-oriented technology is the focus, funding for flow energy development is likely to remain scarce. This is where having a government willing to direct public funding to necessary but modest financial investment-return projects is important.

New water technologies

It appears that Australian policy is to supply water to a population which business has imagined at between 50 million and 200 million[51] by recycling water, desalinating it, and building new dams, in a kind of Saudi or Israeli vision of a twenty-first-century desert economy. All of these schemes which population stabilization and decline would mitigate and avoid, stand to cost the current population socially, physically, and financially, but the benefits (almost exclusively financial) will be focused on only a few. The situation is put to the people as one which will save them from catastrophe, so few question these costs, which they are coached to believe are inevitable.

It would be nice for Australians to have Cuban-style options for reorganizing locally, individually, and cooperatively so that they would not be entirely reliant on the reorganization and repowering of their economy according to the corporate profit motive mediated by semi-privatized government. However, trends are not moving the people's way.

In the long term

Geothermal "hot rocks" technology has not yet progressed to the point where it may be factored into this scenario. This may change. Uranium supply and competition will make substantial refitting of Australia with a nuclear-based system problematic.[52] The nuclear industry plan appears to be for closed-cycle breeder reactors to be built next to conventional reactors and to take over their work reprocessing nuclear wastes and using thorium as uranium depletes. Such a plan will require sustained international support for research, engineering, and construction, and the broadening of an educational matrix for this technology at technical and university level. If, for technical or economic reasons, this grand plan fails, what are we left with?

Hydropower and water transport

Inland water sources on the mainland, with the exception of the unreliable Murray–Darling River system and the Snowy River, are almost all unable to support fluvial transport or reliable power generation. Tasmania is the only state where fast-flowing rivers make widespread hydropower possible.

On the positive side, if substitute big-energy systems are not found, this will also remove the structures and their beneficiaries which currently present barriers to relocalization and land redistribution. On the other hand, by the time this becomes possible, there may be very little salvageable land left.

On current trends,[53] there is going to be less than half the land (due to degradation mentioned above), more than twice the population, and severe water-shortages with greater climate extremes. Based on the above projected losses of arable land, we might assume a loss of at least half of the 5 percent gain in net primary productivity since 1788. Without surplus production or abundant energy, Australia's agricultural and mineral-based export economy will shrink to nothing. Standard of living and quality of life will be locally determined. Without fossil fuel or a new technology substitute, the continent's capacity to support more than its natural biomass will presumably reduce to near zero, and the human population will shrink, one way or another, perhaps to something like 2.5 percent larger than the aboriginal population before 1788 – that is, *below 1.5 million*. Climate change, however, may even reduce this potential.

There are so many variables. The fertile islands of the remaining land could possibly be boosted using permaculture and terra preta technologies and water-saving and tree-growing plans to reform microclimates. It seems clear that permaculture could rehabilitate many degraded soils in Australia. Finding the secret of terra preta, an ancient technology which maximizes carbon-loaded surfaces for soil microrganisms, once employed by South American Indians and others (for example, Northeastern Thailand)[54] is another area of promising soil research.

Such a population might subsist mainly by hunting indigenous fauna (birds, kangaroos) or herding exotic, imported fauna (like cattle), using draft animals (camels, horses, and cattle), and cultivating, by recycling manure and other wastes, a greatly reduced area of arable land. It would use mostly flow energies, of which (based on pre-World War I practices) the most dominant will probably be wind with some solar, if this technology can be retained, and some biomass, if combustion engines are maintained for limited, specific purposes. These flow energies might add to the productivity of the land, but from that gain should be subtracted land needs for work-animal fodder, and the human built environment.

Although it might be possible to use trains and grid electricity on a limited scale using wind,[55] the distance between cities and the fuel-poor-related fall in population and activity would make social organizing to maintain these options unlikely. There might perhaps be a case for wood-fired rail transport from the major inland food production area (should any of this survive), with secondary distribution by road using draft animals, and coastal shipping.

To clarify: The distances and areas required for traditional agriculture in Australia are so vast that they are generally incomprehensible to those from the Northern hemisphere without science-based ecological knowledge. If we strip off all the fossil-fuel-reliant economic and technical features of twenty-first-century Australian society, we are left with isolated communities in an isolated continent where the desert and rangeland most naturally supports a hunter-gatherer economy, some herding, and some agriculture on the parts currently densely occupied, mostly on the coasts.

Described in 1994 as "one nation: two ecologies,"[56] Australia may become a desert and oasis land, if long-term drying trends continue or increase with climate change. Although the pre-1788 Aboriginal population did not possess draft animals, it is not hard to imagine a Bedouin-like alternative operating in the deserts and rangelands, trading between oases. Between 1838 and the 1890, Afghan Cameleers accompanied by hundreds of their camels were brought to Australia to provide desert transport,[57] forging routes which subsequently became roads.

Strong isolation means that large populations may crash and their complex societies and technologies may actually disappear from memory, leaving survivors with little or no connection to the society of their ancestors. Australia is famously a continent which was geographically virtually isolated for between 65,000 and 40,000 years until coal produced the kind of critical mass in iron and fossil-based capital and population to bridge that isolation and that of so many other countries. Keeping that bridge intact is no sure thing.

NOTES

1. S.M. Newman, *The Growth Lobby and its Absence: The Relationship between the Property Development and Housing Industries and Immigration Policy in Australia and France*, Swinburne University, Victoria, Australia, 2002, chapter 7, http://tinyurl.com/yt58yh.
2. "Australia's Population" (Population Clock), Australian Bureau of Statistics, www.abs.gov.au/, November 5, 2007.
3. Including national "baby bonuses" and misleading information from the Federal Treasurer that the population would not grow without immigration (P. Costello (Federal Treasurer, March 1996 to November 2007), Introduction to *Sydney Morning Herald* Supplement, "Births of a Nation," January 26, 2006), and from State Premiers, such as Steve Bracks of Victoria, that "deaths outnumbered births" (when in fact the opposite was true) (Steve Bracks on the Jon Faine Show, ABC Radio, Australia, August 15, 2005, documented in correspondence by Jill Quirk, President, Sustainable Population Australia, Victorian Branch).
4. Institut National d'Etudes Démographiques (INED), Paris.
5. Agriculture Durable et Conservation des Sols, European Conservation Agriculture Federation, www.ecaf.org.
6. J. Tainter, *The Collapse of Complex Societies*, Cambridge University Press, Cambridge, 1998; 12th reprint, 2005, p. 201.
7. F. Ronsin, *La Population de la France de 1789 à nos jours*, Seuil, Paris, 1997.
8. "France's Role in Exploring Australia's Coastline," www.ambafrance-au.org/article.php3?id_article=475, accessed December 3, 2007.
9. "Il aura fallu deux cent deux ans, et vingt ans de discussions, pour enterrer la vision napoléonienne de la famille. Sa «haine» des secondes noces. Sa répugnance à l'égard du divorce. Sa conception du conjoint, considéré comme étranger à la famille. Ainsi, la loi votée le 23 juin «portant réforme des successions et libéralités», et qui sera applicable dès le 1er janvier 2007, donne un véritable coup de balai sur le Code civil, en modifiant plus de 200 de ses articles inchangés depuis 1804." Sabine Delanglade, "Héritage Tout change," *L'Express*, October 26, 2006, www.lexpress.fr/info/economie/dossier/heritage/dossier.asp.
10. See my chapter, "Nuclear Fission Power Options," this volume; Figure 6, "History and forecast of uranium production," in "Uranium Depletion and Nuclear Power: Are We at Peak Uranium?" www.theoildrum.com/node/2379, accessed December 3, 2007.
11. See Michael Dittmar's chapter, "Fusion Illusions," this volume.
12. P. Burger, "La géothermie, une énergie d'avenir pour la production d'électricité," Electrosuisse, February 2, 2006, http://a3.epfl.ch/SeSo/SOdocuments/2006/Electricite_geothermie_A3_orale.pdf, accessed December 3, 2007.
13. Colin Campbell, "The Caspian Chimera," this volume.
14. The standard workhorses of the SNCF are BB locomotives (1960–1970s vintage still in service) and TER *automoteurs* (some 1950s vintage) relying mostly on diesel fuel.

15. R. Darbéra, "Le Versement Transport est-il la meilleure façon de financer la subvention aux transports collectifs urbains?" *Transports* (Paris), no. 368 (November–December 1994), pp. 374–84.; "Velib à Paris, le vélo en libre accès avec ou sans abonnement, c'est quoi?" www.aquadesign.be/news/article-9490.php, accessed December 3, 2007.
16. Newman, *The Growth Lobby and its Absence*, chapter 7.
17. "La Mission de développement des parcs naturels régionaux," www.parcs-naturels-regionaux.tm.fr/upload/doc_telechargement/grandes/plaquette%20Missions%20PNR.pdf, accessed December 2, 2007.
18. Much of this art functioned as maps of areas of land with markers for water, game, people, and landmarks.
19. Total land stock is 770 million hectares, or 7.7 million square kilometers. Estimates of population range between 150,000 and 300,000, through to 900,000.
20. Joseph B. Birdsell, "Australia: Ecology, Spacing Mechanisms and Adaptive Behaviour in Aboriginal Land Tenure," in *Land Tenure in the Pacific*, ed. Ron Crocombe, Oxford University Press, Melbourne, 1971, pp. 334–61.
21. Jennifer Macey, "Vic Bushfires Uncover Ancient Aboriginal Stone Houses," *The World Today*, February 3, 2006, www.abc.net.au/worldtoday/content/2006/s1561665.htm.
22. "With Federation, the Commonwealth Parliament became dominated by spokesmen for 'White Australia'. In October 1901 legislation was passed prohibiting the introduction of Pacific Islanders after 31 March 1904" (R.F. McKillop, referring to G.C. Bolton, *A Thousand Miles Away: A History of North Queensland to 1920*, Australian National University Press, Canberra, 1972, p. 239, in "Australia's Sugar Industry", Light Railway Research Society of Australia website, www.lrrsa.org.au/LRR_SGRa.htm).
23. The Colonial Sugar Refinery Company aroused similar responses among indigenous Fijians who also objected to blackbirding as well as to the importing of Indian indentured labor: "The Indian Connection," *Frontline*, vol. 17, issue 12 (June 10–23, 2000), www.hinduonnet.com/fline/fl1712/17120130.htm.
24. Newman, *The Growth Lobby and its Absence*.
25. For a discussion of Work Choices policy and issues, see Colin Fenwick, "How Low Can You Go?" *Economic and Labour Relations Review*, vol. 16, no. 2 (2006), www.austlii.edu.au/au/journals/ELRRev/2006/5.html.
26. "Largest Population Increase Ever: ABS," Media Release, September 24, 2007, http://tinyurl.com/22z6kd: "Net overseas migration contributed 54% (162,600 people) to this growth, which was more than the natural increase of 46% (138,100 people or 273,500 births minus 135,400 deaths)." This occurred with confusing changes to statistical methods plus new ease of transfer from temporary to permanent migrant (largely equivalent to European citizenship).
27. Tim Flannery, *The Future Eaters*, chapter 27, Reed Books, London, 1994, pp. 368–9.
28. In the words of Dorothea McKellar's poem, "My Country" (1908).
29. Flannery, *The Future Eaters*, p. 114.
30. Ibid., pp. 92–3.

31. B. Foran and F. Poldy, *Future Dilemmas: Options to 2050 for Australia's Population, Technology, Resources and Environment*. Report to the Department of Immigration and Multicultural and Indigenous Affairs by CSIRO Sustainable Ecosystems, October 2002, p. 130, www.cse.csiro.au/publications/2002/fulldilemmasreport02-01.pdf.
32. Macey, "Vic Bushfires Uncover Ancient Aboriginal Stone Houses."
33. Net productivity is measured by carbon gained through photosynthesis and carbon lost through plant respiration. Net primary productivity averages 0.96 Gt of carbon annually for Australia. Nearly 60 Gt of the total continental carbon is stored as plant biomass (45 percent) and soil carbon (55 percent) (*Australian National Resources Atlas*, http://audit.ea.gov.au/anra/agriculture/docs/national/Agriculture_Landscape.html).
34. Carbon, organic nitrogen, and organic phosphorous have nearly doubled since 1788. Some of this is now counterproductive (*Australian National Resources Atlas*).
35. For the lower figure: S. Siedlecky and D. Wyndham, *Populate and Perish, Australian Women Fight for Birth Control*, Allen and Unwin, Sydney, 1990, p. 142. For the higher figure: James Jupp, ed., *The Australian People: An Encyclopedia of the Nation, its People and their Origins*, Angus and Robertson, North Ryde, NSW, 1988, p. 148 (the higher figure refers to estimates by Noel Buttlin in 1983). Cook and Joseph Banks commented on the low density of the Aboriginal population and Banks correctly inferred this to be a consequence of the low fertility of the land. Thomas Malthus wrote in his second volume that he was inspired by these comments to write his first volume on population (T. Malthus, *An Essay on the Principle of Population and a Summary View of the Principle of Population*, Penguin Classics, London, 1985, p. 251.
36. Foran and Poldy, *Future Dilemmas*, p. 125.
37. Ibid; such as legume pastures without balancing lime applications and nitrogenous fertilizers.
38. F. Morgan (Director), *The Power of Community: How Cuba Survived Peak Oil*, Alchemy House Productions Inc. and Community Solution, 2006; D. Pfeiffer, "Cuba – A Hope," www.fromthewilderness.com/free/ww3/120103_korea_2.html.
39. J. Skatssoon, "'Censorship' Just Tip of Iceberg," ABC Science Online www.abc.net.au/science/news/stories/s1569599.htm, accessed November 20, 2007.
40. www.abareconomics.com/interactive/energy_july07/index.html.
41. A petajoule (PJ) is 10^{15} (one quadrillion) joules.
42. "Nuclear Energy Prospects in Australia," Briefing Paper No. 44, September 2007, www.uic.com.au/nip44.htm.
43. Ian Hore-Lacey (Director – Information), Australian Uranium Association, provided these petajoule estimates (email correspondence, November 30, 2007).
44. "Primary Energy Consumption in Australia by Fuel 2005–2006," www.abareconomics.com/publications_html/energy/energy_07/energy_update_07.pdf.
45. Using centrifugal separation and depending on quality of uranium ore; www.uic.com.au/nip57.htm.

46. Newman, *The Growth Lobby and its Absence*, chapter 6.
47. VAMPIRE index: J. Dodson and N. Sipe, "Oil Vulnerability in the Australian City: Assessing Socio-economic Risks from Higher Urban Fuel Prices," *Urban Studies*, vol. 44, no. 1 (2007), pp. 37–62.
48. Jennifer Marohasy, "Toowoomba Votes Against Water Recycling," www.jennifermarohasy.com/blog/archives/001511.html; "Your Water Your Say," www.yourwateryoursay.org/; "Communities Against the Tunnels Forum," www.notunnels.org/; "Blue Wedges Against Deepening Port Phillip Bay," www.bluewedges.org/index.php?page=home.
49. See Alice Friedemann's chapter, "Peak Soil," this volume.
50. As Andrew McKillop suggests in his chapter "No Choice but International Energy Transition," this volume.
51. Newman, *The Growth Lobby and its Absence*, chapter 6.
52. See my chapter, "Nuclear Fission Power Options," this volume, on uranium supply.
53. Foran and Poldy, *Future Dilemmas*, p. 125.
54. See Chapter 16, this volume.
55. J. Coulter, "Size, Cost and Timing of Change." Paper to the National Conference of Sustainable Population Australia, University of Adelaide, 2002.
56. Title of a carrying capacity report in 1994, resulting from an interparliamentary inquiry chaired by Barry Jones.
57. "Australia's Muslim Cameleer Heritage," http://tinyurl.com/2afed7.

21

North Korea: The Limits of Fossil Energy-Based Agricultural Systems

What North Korea Tells Us About Our Future

Antony Boys

INTRODUCTION

In the 1990s, the Democratic People's Republic of Korea (DPRK)'s agricultural system collapsed and the people began to starve. Inadequate energy supplies were the immediate cause. Loss of soil fertility through economic mismanagement and the inability to trade on the world market for energy made modern agriculture impossible. In consequence the DPRK could no longer provide sufficient food for its population through domestic production. This chapter shows how commercial energy shortages directly influence the productivity of modern agricultural systems.

The fragility of modern industrial (chemical and fossil fuel-based) agricultural systems is highlighted here. If trade is disturbed, interrupting energy (fossil resource) supply over an extended period, agricultural systems are liable to fail, causing food shortages. Conclusions from the discussion of North Korea's recent tragic experience in this chapter also furnished the basis for the next chapter in this part, about the prospects for Japan in a period of oil decline.

OVERVIEW OF THE DPRK ENERGY CRISIS

The DPRK relies heavily on indigenous sources of power, predominantly coal and hydropower, and has no known reserves of oil or natural gas. Since the end of the Cold War, chronic shortages have developed for all forms of modern energy supply, with petroleum products, coal, and electricity all reduced by more than 50 percent after 1990. These shortages have affected all sectors of the economy, especially transportation, industry, and agriculture. The North Korean energy crisis results from the loss of subsidized Soviet oil imports, failure to maintain and modernize energy infrastructure, the impacts of natural disasters, and inefficiency in energy production and end use.[1]

During the Cold War, the DPRK received heavily subsidized oil supplies from the Soviet Union. In 1990, crude oil imports amounting to about 2.5 million tonnes were delivered from the USSR, the People's Republic of China (PRC), and Iran, and a further 600,000 tonnes of refined petroleum products such as diesel and gasoline from China. Crude oil delivered by tanker was refined at Raijin, and at the terminus of a pipeline from China.[2] With the collapse of the USSR in 1990, subsidized oil supplies to the DPRK and other former client states, such as Cuba, were curtailed. The DPRK could not pay world market prices for oil, and imports from Russia and the Middle East soon fell by 90 percent. China became the main supplier of oil to the DPRK, oil imports in 1996 being around 40 percent of their 1990 level (around 1.1 million tonnes).[3] Imported oil, limited to non-substitutable uses such as motor gasoline, diesel, and jet fuel, accounted for about 6 percent of primary energy consumption.[4]

In the same period, coal supply fell 50 percent from around 34 million tonnes in 1990 to around 17 million tonnes in 1996. Electricity supply fell 52 percent from around 46 billion kilowatt hours (kWh) in 1990 to around 24 billion kWh in 1996. Estimated total consumption of commercial energy in the DPRK fell by about 51 percent. Biomass use rose about 8 percent from around 22.3 million tonnes to around 24 million tonnes. The capacity of the main forms of transportation for goods in the DPRK, electric and diesel trains, and diesel trucks probably fell to 40 percent of its 1990 value by 1996. Iron and steel production shrank to 36 percent of 1990 levels by 1996.[5]

In 1997, coal accounted for more than 80 percent of primary energy consumption and hydropower more than 10 percent, with hydroelectric power plants generating about 65 percent of North Korea's electricity, and coal-fired thermal plants about 35 percent.[6]

To summarize, the shortage of one relatively small but key element in the DPRK energy mix, imported oil, appears to have set off systemic shockwaves throughout the DPRK economy, causing a more than 50 percent drop in all economic activity, including basic food production (see Table 21.1). *What may well have occurred (although there is little direct evidence to back this up) is that much of the imported oil was being used to operate coal mines. When that oil ceased to be imported, the operation of coal mines became immediately problematical, causing the production of coal to plummet to half the 1990 level.*

Table 21.1 DPRK energy situation 1990 and 1996

	Units	*1990*	*1996*	*1990 (PJ)*	*1996 (PJ)*	*1996/ 1990 %*	*% total 1990*	*% total 1996*
Crude oil	million tonnes	2.5	1.1	114.3	50.3	44.0	9.9	8.8
Refined products	million tonnes	0.6	0?	27.1	0.0	0.0	2.4	0.0
Coal	million tonnes	34.0	17.0	928.2	464.1	50.0	80.8	81.3
Hydroelectricity	billion kWh	22.0	15.6	79.2	56.2	70.9	6.9	9.8
TOTAL				1148.8	570.5	49.7	100.0	100.0
Electricity (total)	billion kWh	46.0	24.0	165.6	86.4	52.2		

Notes: "Electricity (total)" here includes electricity generated from coal, and thus is shown for reference and not included in the energy total.

Generation of hydroelectricity in 1996 is calculated as being 65 percent of the total of generated electricity.

Conversion rates used to convert between differing energy units (www2.dti.gov.uk/epa/annexa.pdf) were:

Electricity: 1 kWh = 0.0036 GJ
Coal: 1 tonne = 27.3 GJ
Crude oil: 1,181 liters/tonne
Crude oil: 45.7 GJ/tonne
Crude oil: 1 barrel = 159 liters
Crude oil: 1 barrel: 0.1346 tonnes = 6.15 GJ

Sources: James H. Williams, David von Hippel, and Peter Hayes, "Fuel and Famine: Rural Energy Crisis in the Democratic People's Republic of Korea," Institute on Global Conflict and Cooperation, IGCC Policy Papers, Paper No. 46, March 2000, http://repositories.cdlib.org/igcc/PP/pp46.

Generation of hydroelectricity in 1990 is estimated from http://www.eia.doe.gov/emeu/cabs/nkorea.html.

THE COLLAPSE OF MODERN AGRICULTURE IN THE DPRK

The collapse of motive power: the substitution of human and animal labor for machines and commercial energy

Agriculture in the DPRK has been organized as cooperative and state farms and has concentrated mainly on the production of rice and maize. Since the 1950s, modernization of agriculture has been carried out through the promotion of irrigation, electrification, chemicalization (fertilizers, pesticides, herbicides, and so on) and mechanization. The 1970s and early 1980s saw fruition of these efforts, irrigation reaching 70 percent or more of the cultivated land by 1970; a total of 75,000 tractors as well as transplanters, threshers, trucks and other farm machinery were provided; rural electrification covered all rural areas by 1968; and fertilizers and other chemicals became available in large quantities.[7]

Since the early 1990s, resource constraints brought about by an ailing economy, as described above, made provision of previous levels of inputs difficult and made it impossible to maintain land and labor productivity which depended on the high input-based agricultural system. Much of the agricultural machinery and equipment was purchased in the 1950s and 1960s and is now obsolete and/or in a very poor state of repair due to the inability to purchase or manufacture sufficient spare parts. It is estimated that 80 percent of the 3 million horsepower (hp) (2,200 megawatts, MW) of motorized capacity in the DPRK agricultural sector was inoperable at the end of 1998.[8]

Fuel to operate the machinery for critical mechanized operations has also become exceedingly scarce. Based on typical consumption rates of 110–140 liters per hectare per year (L/ha/yr) for rice and maize, annual fuel requirements for DPRK farms have been estimated at 140,000 tons of petroleum products, mostly diesel fuel. In 1990, the DPRK agriculture is estimated to have used at least 120,000 tonnes of diesel fuel. Following the crisis, this declined to between 25,000 and 35,000 tonnes in 1996 and 1997. Total diesel supplies fell from 750,000 tonnes in 1990 to around 300,000 tonnes per year after 1996. Military allocations have remained almost steady at 160,000 tonnes, leaving only 140,000 (roughly equal to the annual requirements for agriculture alone) for all other allocations, leaving little diesel fuel for agriculture.[9]

During field visits, the October 1998 Food and Agriculture Organization/World Food Program (FAO/WFP) Mission saw a large proportion of tractors, transplanters, trucks, and other farm machinery lying unused and unusable, as well as harvested paddy left in the fields in piles for three weeks or more, resulting in large post-harvest losses. Decreasing ability to carry out mechanized operations (including the pumping of water for irrigation), as well as lack of chemical inputs, was clearly contributing to reduced yields and increased harvesting and post-harvest losses.[10]

Energy has been a critical factor in the operation of the irrigation system of the DPRK. Water is pumped into the main canals and reservoirs, but fuel scarcity has made it impossible to guarantee timely supplies of water to the irrigation system. Around 1 million ha of rice, maize, and other crops are irrigated through more than 10,000 km of canals and pipes by more than 30,000 pumping stations, mostly electrical. UN irrigation experts estimate the electricity requirement for the DPRK averages 1,200 kWh/ha/yr, representing an annual national requirement of 1.2 billion kWh. In 1996, this figure was 0.9

billion kWh, a 300 million kWh (25 percent) shortfall for irrigation pumping. With 24 billion kWh of electricity generated in 1996, it might appear that the DPKR could reassign power from other sectors to agriculture. However, over half of irrigation pumping takes place in May, when peak pumping power demand is at least 900 MW. Total national generating capacity in 1996 was 4.7 GW. If the average capacity factor was 0.65, the average generating capacity online would have been 3.1 GW, from which must be subtracted about 19 percent for transmission and distribution losses. Peak irrigation pumping demand would thus represent over 30 percent of the DPRK's total generating capacity! The 25 percent shortfall in electricity for irrigation pumping would have led to a similar shortfall in irrigation water provided to crops, resulting in declining crop yields.[11] The condition of canals and pumping stations has also been severely affected by natural calamities, while the pumping stations and steel pipes used in the system have suffered from a lack of spare parts and poor maintenance. The breakdown of the irrigation system due to lack of spare parts and electricity shortages has been a major factor in the severe deterioration in rice production.[12]

Total agricultural electricity requirement is estimated at around 1.7 billion kWh, some 460 million kWh being required to operate electrical machinery other than irrigation pumps. A further 1.2 billion kWh is required for domestic, public and commercial uses in rural areas, bringing the grand total of electricity required in rural DPRK to 2.9 billion kWh/yr. In the late 1990s, the actual supply was estimated at 1.9 billion kWh, a shortfall of roughly 1 billion kWh, agricultural uses receiving around 1.3 billion kWh and other uses being reduced by half to 0.6 billion kWh.[13] Household coal consumption for cooking, heating, and preparing animal feed declined on average 40 percent. In areas with limited access to biomass as a substitute fuel, there were undoubtedly serious impacts on health and quality of life; for example, due to the inability to boil water. Relief workers at the time reported significant health effects from waterborne diseases.

In order to maintain agricultural production despite the declining availability of machinery and equipment, it has been increasingly necessary to use manual labor and work animals.[14] In the late 1980s, approximately 25 percent of the civilian workforce was engaged in agriculture, and by the mid-1990s this rose to about 36 percent, or around 1 million people.[15] The additional labor required to compensate for the lack of mechanized inputs is conservatively estimated at a minimum of 300 million person-hours per year.[16]

This could easily be supplied by the extra 1 million people in the rural workforce. The June 1998 FAO/WFP Mission observed large numbers of the non-agricultural population, including school children, assisting with land preparation, planting and weeding activities.[17] In October 1998, the Mission observed people carrying bundles of harvested paddy on their backs to threshing stations and school children collecting grain around the harvested crop bundles in the fields.[18]

Nevertheless, it has proved impossible to perform all operations previously carried out by machinery simply by use of manual labor or work animals (for example, plowing, pumping of water for irrigation, harvesting, threshing, and so on). Thus the timely and effective scheduling of farm operations, particularly at peak times around harvesting and planting of double crops, has been severely affected, reducing productivity and increasing post-harvest losses.[19] Work animals have been increasingly substituted for farm machinery, but this option has been limited due to low numbers of animals available. Animal health is poor due to lack of feed, because of competition with humans for arable land.[20]

Biomass fuels are mainly used for household cooking and heating. In 1990, rural sector biomass fuel consumption was estimated at 22.7 million tonnes, but by 1996 this had risen by an estimated 1.3 million tonnes per year to make up for shortfalls in coal and other fuels. The problem with a rise in biomass fuel consumption is that it takes biomass away from other potential uses, such as for animal fodder and compost, and this in turn has adverse effects on food supplies. By reducing ground cover, disrupting habitats, and increasing soil erosion and siltation, rural ecosystems such as forests, streams, and croplands are also adversely affected by increased biomass use.[21]

The collapse of soil fertility: mining the soil

Modern agriculture relies on steady inputs of inorganic chemical fertilizers. For grain crops under North Korean soil and growing conditions, the amount required is 400–500 kilograms per hectare (kg/ha) of the basic macronutrients nitrogen, phosphate, and potassium (NPK). UN and DPRK agricultural experts estimate the total North Korean requirement at 700,000 tonnes/year (NPK).[22] The actual bulk amount of fertilizer required to fulfill this demand would be of the order of 1.5–2.5 million tonnes year, since the nutrient content of specific fertilizers differs (for example, urea contains more than twice the amount of nitrogen than ammonium phosphate per unit weight). Historically, the DPRK had manufactured 80–90 percent of its fertilizer requirement, importing about 20 percent of its phosphate fertilizer, and all of potassium fertilizer.[23]

The calamitous decline in fertilizer production is a result of fertilizer factories being out of operation or operating at minimal levels. This was reported to be at least partly due to the poor condition of Soviet-built plants, natural disasters having perhaps caused some of the damage. Prior to 1990, the DPRK is said to have operated three fertilizer factories, capable of a total annual production of over 400,000 tonnes of nitrogen nutrient. This would have provided for self-sufficiency if the plants had been able to run at capacity.[24] The important nitrogen fertilizer plant at Hamhung had apparently been inoperable since at least 1994, and the DPRK government had requested international assistance to refurbish the plant. Notwithstanding problems of damage or disrepair, the energy crisis would have had several important effects on fertilizer production. The DPRK nitrogen fertilizer production process uses coal as both energy source and chemical feedstock. Coal is hydrogenated to produce hydrogen, which is then reacted with nitrogen from the air to produce ammonia, the basic chemical starting point for nitrogen fertilizers, through the Haber-Bosch synthesis. To produce 700,000 tonnes of nitrogen per year would require an estimated 1.5–2.0 million standard tonnes of coal (approximately 60:40 fuel:feedstock). This would have accounted for around 10 percent of the annual coal supply in 1996, creating severe competition with other high-priority uses. Further, transporting up to 2 million tonnes of coal would represent a serious strain on the transportation system; especially the railways, which were already suffering from severe electricity shortages. Transportation limitations would also have made the shipping of 1.5–2.5 million tonnes of fertilizer from factories to farms quite problematical. It was estimated in 2000 that DPRK freight shipments by ship, rail, and truck had declined by 55 percent, 60 percent, and 75 percent, respectively, since 1990. Williams et al. point out that for these reasons, even if the DPRK's fertilizer plants were refurbished or rebuilt, energy shortages would continue to pose a serious constraint on domestic fertilizer supply.[25]

Due to the fertilizer shortage, DPKR agriculture operated at 20–30 percent of normal levels of soil nutrient inputs between 1996 and 2002 (see Figure 21.1). This shortfall is the largest single contributor to reduced crop yields, and thus to food shortages. Soils in the DPRK are being heavily mined of nutrients, as more nutrients are being extracted than replaced. According to the DPRK Ministry of Agriculture, 244,512 nutrient tonnes of fertilizer were used in agriculture in the DPRK during 2003 (Figure 21.1), compared with only 189,000 tonnes in the previous year. Most was provided as humanitarian assistance by the Republic of Korea, the European Union, the FAO and various non-governmental organizations. Just over 32,000 nutrient tonnes were produced in the DPRK, and 37,706 nutrient tonnes were imported commercially. Of the total used, 68 percent was nitrogen, mostly in the form of urea; 15.5 percent was phosphorus; and 16.5 percent was potassium.[26]

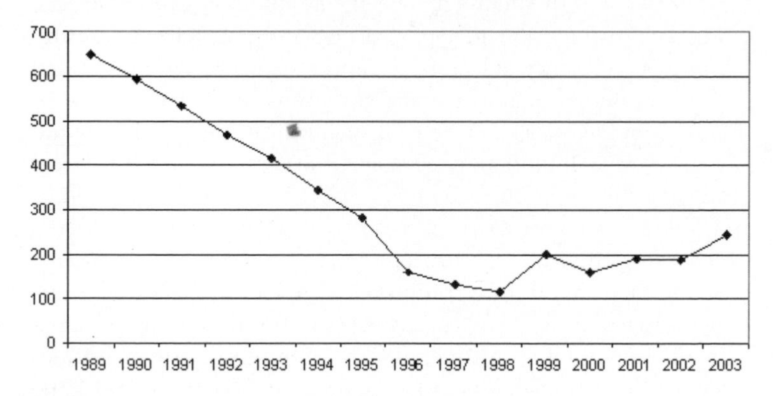

Figure 21.1 DPRK consumption of nitrogen, phosphorus, and potassium, 1990–2003 (1,000 nutrient tonnes)

Source: FAO Global Information and Early Warning System on Food and Agriculture, World Food Program Special Report, *FAO/WFP Crop and Food Supply Assessment, Mission to the Democratic People's Republic of Korea*, October 30, 2003, FAO Corporate Document Repository, www.fao.org/docrep/006/j0741e/j0741e00.HTM, accessed March 7, 2008.

The modern system of intensive agriculture introduced in the DPRK between the 1950s and 1980s enabled the continuous production, including double-cropping, of cereals, but has resulted in soils highly depleted in natural nutrients. Rotational systems including fallowing or planting with leguminous crops have been abandoned, as, largely, has the practice of using organic fertilizers. These systems can, of course, be reintroduced and would be expected to raise yields as they help to revive soil fertility. These are, however, systemically complex, long-term measures which can only be reintroduced with careful consideration for the overall productive capacity of the DPRK's agriculture, the ability to import substantial quantities of food, existence and location of relatively large numbers of livestock, availability of surplus organic matter (for example, not in competition with livestock feed or biofuel production), and so on. For the time being the DPRK remains heavily reliant for its food supply upon chemical fertilizers, which must either be manufactured at considerable energy and raw material cost or imported.[27]

The collapse of cereal production

The staple foods of the DPRK are rice and maize, and these crops are the main food grains produced in the country, accounting for about 1.2 million ha of the country's total arable land of about 2 million ha. Rice is grown predominantly in the southern plains, while maize is grown generally on sloping ground. Rice is transplanted from mid-May to early June, harvested from late September to October, and is almost totally irrigated. Maize is largely rain-fed, planting being carried out from mid-April to early May, harvesting coming between the end of August and mid-September. Irregular or poor rainfall will therefore affect maize more than rice. Whereas the 1997 drought devastated maize production, rice production has been adversely affected by breakdowns in the irrigation system, as mentioned above.[28] There has been a shift in recent years away from the cultivation of maize in favor of potatoes on low yielding and vulnerable (easily eroded or degraded) sloping land. This shift has been in the region of 110,000 ha. Double-cropping of maize and potatoes has also been reported; the area planted being around 77,000 ha.[29]

Statistics showing approximate cultivated land areas and food production derived from FAO reports follow. Table 21.2 shows land classification of arable land cultivated to paddy and maize in the DPRK in 1998 and gives a rough idea of total areas under these two main crops. Table 21.3 shows estimated cereal production and availability for the 2004/05 marketing year and total arable land area available for food crops, approximately 1.4 million ha.

Figure 21.2 gives approximate production figures for cereals (as defined in Table 21.3) for the crop years 1989/90, and from 1995/96 to 2004/05 (the figures in Figure 21.2 correspond to those shown in Figure 21.3). It is clear from the graphs just how much staple grain production declined in the DPRK, from being more than 1 million tons in surplus in the late 1980s to being more than 1 million tons in deficit in the mid-1990s. A comparison of Figure 21.1 (fertilizer availability) with Figure 21.2 (cereal production) suggests that there is a strong correspondence between fertilizer availability and cereal

Table 21.2 Classification of rice and maize areas cultivated 1998

Land type	*Rice*		*Maize*		*Total area*
	Area (ha)	*%*	*Area (ha)*	*%*	
Good: Class I	188,000	32.4	202,000	32.1	390,000
Moderate: Class II	195,000	33.6	195,000	31.5	390,000
Poor: Class III	197,000	34.0	229,000	36.4	426,000
Total	580,000	100	626,000	100	1,206,000

Source: FAO, 1998/11, Section 3.7.1.

Notes:

Class I lands: Flat and/or leveled, good soil, irrigated, farm machinery and equipment routinely used.

Class II lands: Flat or undulated (gradient 0–10 degrees), irrigated or unirrigated, good to moderate soil, farm machinery and equipment often used.

Class III lands: Sloped and hilly (flat for paddy), no irrigation except for paddy, unsatisfactory soil, farm machinery or equipment not normally used.

Table 21.3 DPRK cereal area and production, 2004/05

Crop	*Area (1,000 ha)*	*Yield (tonnes/ha)*	*Production (1,000 tonnes)*	*Total cereal area* (1,000 ha)
Rice (unhulled) [A]	583	4.10	2,370	583
Maize [B]	495	3.50	1,727	495
Potato [C]	189	10.90	2,054	189
Wheat and Barley, double-cropped [D]	102	2.26	231	102
Other cereals [E]	60	2.00	119	60
Rice in milled equivalent[a] [F]			1,540	1,429
Potato in cereal equivalent[b] [G]			513	
Total Production (cereal equivalent) [B + D + E + F + G]			4,130	

[a] Milling rate of 65 percent.

[b] Potato to cereal equivalent 25 percent (4:1).

Source: FAO, 2004/11, Section 4.4.

production. It appears that in the mid to late 1990s, food production in the DPRK fell to about 45 percent of that of the 1980s (approximately the same level as the availability of commercial energy resources), and that it cannot regain its former level without the revival of substantial inputs of chemical fertilizer and the restoration of the former mechanical capacity of DPRK agriculture. This is borne out by the fact that even in 2004/05 the grain production shortfall was nearly 900,000 tonnes. In late March 2007 the DPRK admitted for the first time to food shortages of 1 million tonnes.[30]

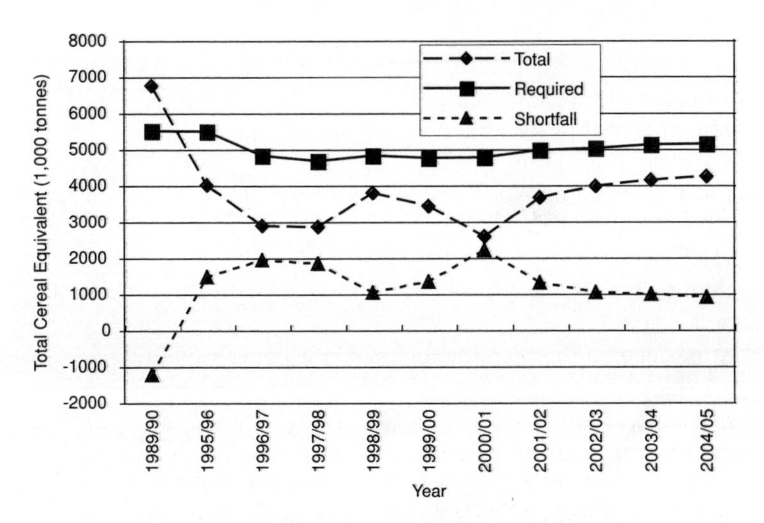

Figure 21.2 Total cereal equivalent production, requirement, and import requirement ("shortfall") of DPRK 1989–2004

Notes:
The value for total production for 2004/05 (4,235,000 tonnes) is obtained by adding a further 105,000 tonnes for gardens and slopes to the final figure (4,130,000 tonnes) in Table 21.3.
Total = Total production of maize, milled rice (and potatoes converted to cereal equivalent from 1999 onward).
Requirement = cereal required for food and feed use, seed requirement, other uses, and post-harvest losses.
Shortfall = import requirement to cover cereal production shortfall.
2003/04 Requirement figure is estimated.
Sources: UN Special Reports on FAO/WFP Missions to the DPRK, June 1998 to November 2004.

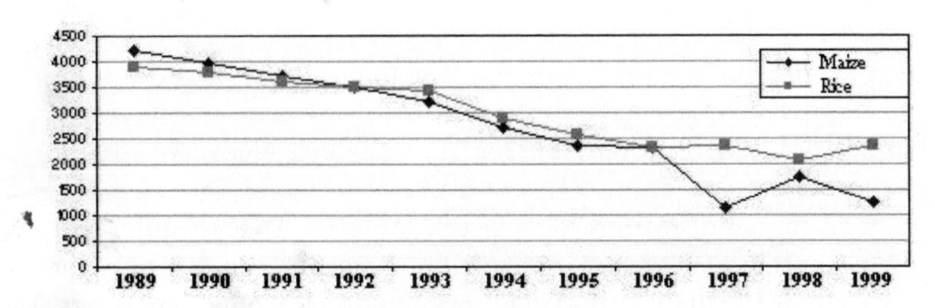

Figure 21.3 Rice and maize production in DPRK 1989–99 (1,000 tonnes)

Note: Value for rice is "unmilled."

Source: FAO Global Information and Early Warning System on Food and Agriculture, World Food Program Special Report, *FAO/WFP Crop and Food Supply Assessment, Mission to the Democratic People's Republic of Korea*, November 8, 1999, FAO Corporate Document Repository, www.fao.org/docrep/004/x3691e/x3691e00.htm, accessed March 7, 2008.

CONCLUSION

The eventual answer to the DPRK food production problem must be to attempt the transition to intensive organic agriculture; low levels of use of commercial energy sources and chemicals, a tight recycling of nutrients in combination with other methods of maintaining soil fertility, such as rotational systems, diversification of crops, and the development (return to!) integrated crop and livestock production systems. This has been attempted fairly successfully in Cuba in recent years.[31]

More generally, we can say the following concerning the *cause* of the DPRK food crisis:

Inability or unwillingness to participate in the global trading economy can cause difficulties in maintaining levels of commercial inputs necessary for continuous operation of a modern food-producing agricultural system.

The experience of the DPRK, and perhaps Cuba, points to several *closely interlinked lessons* that need to be learned by countries which currently operate a modern industrialized agricultural system based on commercial chemical and energy inputs. Agriculture has now become simply one adjunct of the overall economic-industrial matrix of the human global social-economic entity. This matrix is a highly complex web of financial and industrial relationships backed up by fairly precisely timed operations, such as transport of raw materials, fuel, components, and so on. Adjuncts to the matrix are therefore sensitive to disruptions and other irregularities. Thus the modern agricultural system can very quickly get into deep trouble if we do not have the ability to:

1. fuel, maintain, repair, and replace agricultural and distribution-related machinery and infrastructure (trucks, tractors, transplanters, harvesters, irrigation pumps, fuel and chemical delivery systems, and so on)
2. fuel, maintain, repair, and replace factories and factory equipment for the manufacture of vital agricultural machinery and inputs; for example, regularly replaced items such as spark plugs and filters, spare parts, fertilizers, herbicides, pesticides, plastic sheeting, and so on
3. ensure trade and transportation arrangements for steady supplies of fuel, raw materials, and feedstocks for agricultural operations and inputs, such as petroleum, natural gas, coal, potassium and phosphorus minerals, and so on.

Again, the final answer is to convert to *low-input, yet land and labor intensive, organic farming*. Crucially, this would require perhaps a ten- to twenty-year transition period, something the DPRK has not had the luxury of.

As a final general statement, it can be said that once a country takes the decision to abandon traditional agriculture and switch to a modern agricultural system (a mechanized system making use of commercial chemicals and fuels), then in order to maintain food production levels it is essential to ensure that levels of fuel and other inputs are maintained, and that machinery and equipment is kept in good working order. Shortages of fossil resources (oil, natural gas, and coal) can result in productivity collapses when soils are mined, and eventually destroyed, due to crop production without replacement of essential nutrients, and where agricultural machinery and equipment can no longer be kept operational because of lack of fuel and maintenance.

A transition to organic and/or traditional and sustainable forms of agriculture is not easily carried out quickly (for instance, due to lack of livestock and sufficient numbers of farmers with the requisite knowledge and skills). Meanwhile, the population must be fed; a population that has ballooned on food produced by the modern industrial agricultural system that has been built up thanks to fossil resources.[32] This is now the paradoxical complex of problems faced by most of the world, including the great food-producing areas of North America, Europe, South America, and Oceania; the central element that has made high agricultural productivity possible, oil, is at the same time responsible for the deterioration of our most important resource, soil fertility. How do we now ensure the maintenance of that high productivity in the face of future energy shortages? The end of cheap and abundant oil and other fossil resources, symbolized by the peak in conventional oil extraction, probably occurring during this first decade of the twenty-first century, means the end of our current methods of food production, and thus it possibly spells the end of advanced industrial society as we know it. The DPRK is an exceptional case only in the sense that, due to political miscalculation and mismanagement of its economy, it has manifested these symptoms before fossil resource shortage becomes a serious concern for most of the world.

NOTES

1. James H. Williams, David von Hippel, and Peter Hayes, "Fuel and Famine: Rural Energy Crisis in the Democratic People's Republic of Korea," Institute on Global Conflict and Cooperation, IGCC Policy Papers, Paper No. 46, March 2000, http://repositories.cdlib.org/igcc/PP/pp46, p. 1.
2. Ibid., p. 4.
3. Ibid., p. 5.
4. US Energy Information Administration (EIA), www.eia.doe.gov/emeu/cabs/nkorea.html.
5. Williams et al., "Fuel and Famine," pp. 5–6.
6. EIA, www.eia.doe.gov/emeu/cabs/nkorea.html.
7. FAO, Global Information and Early Warning System on Food and Agriculture, World Food Programme Special Report, *FAO/WFP Crop and Food Supply Assessment, Mission to the Democratic People's Republic of Korea* [henceforth *FAO*], 1998–2004, 1998/11, Section 2, www.fao.org/giews/english/alert/index.htm, accessed May 10, 2007.
8. Ibid.; 1999/11, Section 1; Williams et al., "Fuel and Famine," p. 8.
9. Williams et al., "Fuel and Famine," p. 8.
10. *FAO*, 1998/11, Section 3.5.
11. Williams et al., "Fuel and Famine," pp. 9–11.
12. *FAO*, 1998/11, Section 3.2.
13. Williams et al., "Fuel and Famine," pp. 10–11.
14. *FAO*, 1998/11, Section 2.
15. Nicholas Eberstadt, *The End of North Korea*, AEI Press, Washington, 1999, p. 67; Australia, "Department of Foreign Affairs and Trade (DFAT) – Country Information – Democratic People's Republic of Korea – Economic Overview," www.dfat.gov.au/geo/dprk/dprk_brief_economic.html.
16. Williams et al., "Fuel and Famine," p. 12.
17. *FAO*, 1998/6, Section 2.2.
18. *FAO*, 1998/11, Section 4.1.
19. *FAO*, 1999/11, Section 2.
20. *FAO*, 1998/11, Section 3.5.
21. Williams, "Fuel and Famine," p. 12.
22. *FAO*, 1998/6, Section 2.3; 1999/6, Section 2.2; 1999/11, Section 3.2.
23. Williams et al., "Fuel and Famine," pp. 7–8.
24. *FAO*, 1998/6, Section 2.3.
25. Williams et al., "Fuel and Famine," p. 13.
26. *FAO*, 2003/10, Section 4.3.
27. *FAO*, 1999/6, Section 2.2.
28. *FAO*, 1998/11, Section 3.
29. *FAO*, 1999/6, Section 2.2; 1999/11, Section 3.4.
30. Lindsay Beck, "North Korea Facing 1 Million Tonne Food Shortage – WFP," Reuters, March 26, 2007, www.alertnet.org/thenews/newsdesk/PEK169861.htm, accessed May 10, 2007.
31. Peter Rosset and Medea Benjamin, eds., *The Greening of the Revolution – Cuba's Experiment with Organic Agriculture*, Ocean Press, Australia, 1994.
32. Vaclav Smil, "Global Population and the Nitrogen Cycle," *Scientific American* (July 1997), pp. 76–81.

22
How Will Japan Feed Itself Without Fossil Energy?[1]

Antony Boys

This chapter is about the future of energy and society in Japan without today's plentiful supplies of fossil energy resources, especially of petroleum. It based its premise on the author's study of North Korea's battle to survive in the period following the withdrawal of access to cheap Russian oil, also in this volume.

Since we cannot know the future directly, we can only subject current trends to different hypothetical circumstances. It is not possible to make actual *predictions*. What we can do is look at how people lived before the fossil energy revolution, the response of politically isolated countries like North Korea and Cuba to local fossil energy scarcity following the fall of the USSR, and apply our current technological and resource knowledge to the problem of impending shortages of fossil energy for different countries and regions.

We can also learn from imaginative and sociological works.

A writer on the Edo Period (1603–1867) economy and lifestyle, Eisuke Ishikawa,[2] very effectively evokes, in a novel, the lifestyle of Japan in 2050. He obliquely relates (through the memories of an old man living at that time) how Japan reached that point. Reading between the lines, the Japanese economy slowly begins to wind down after about 2012. Food and energy imports severely reduce around 2015 and are non-existent by 2020, by which time the big cities are essentially empty. There follows a very difficult period when there is obviously some starvation (to which Ishikawa hardly refers at all) and by 2050 the population of Japan reaches about 65 million people, of whom 99 percent are living rural, farming lifestyles in independent communities. Though politically very different, the lifestyle has effectively "returned" to a lèisurely and idyllic-sounding Edo Period. 65 million people is roughly the population of Japan in 1930, and the population level we would expect to see between 2070 and 2080 under the low population scenario by Japan's National Institute of Population and Social Security Research (IPSSR – see below).

Ishikawa's scenario could be termed a "slow crash." Ishikawa assumes (not unreasonably) that when the economy begins to shrink, the population at large does not understand what is happening and the government makes no special effort to alleviate the oncoming crisis. This is made clear by the memories of the old man in the story. As the crisis deepens there is a window of opportunity for people with links to the countryside (many Japanese city-dwellers still have fairly strong links with their relatives in the rural areas) to abandon the cities and set up home with their extended families. Those who are lucky enough to do so under favorable circumstances (and the village in Ishikawa's story is very well-off in this respect) have relatively little trouble defending their livelihoods and making the transition to the new lifestyle. Others – hinted at, but not mentioned directly in the novel – do not do so well.

In the Japan that exists outside the novel, under the IPSSR low population scenario the population would be 92 million in 2050 (but 65 million between 2070 and 2080). To reach 65 million in 2050, a "drop" in the population of perhaps 20–30 million people would have to take place in the 2020s.

In today's Japan, population is concentrated in the cities, where the great part of fossil energy-based employment has taken place in the last 150 years. Most Japanese do not live where food is produced or where the farms could be located in the future. As Ishikawa suggests, there may be considerable population movement, but it will not be easy for people to know where and when to go. Contemporary Japanese lack farming experience, skills or knowledge. Daily hard physical labor has become unfamiliar to many. It may take decades to develop a highly productive agricultural workforce. There may be a shortage of teachers. There may not be enough fuel for machines; it may not be possible to keep the machines in good repair. Farm animals may be rare. Much of the hard, physical work may have to be done by human labor. But, with a little foresight, the worst problems could perhaps be lessened.

By informing ourselves of potential future problems we may also identify potential new options, such as what kind of society we want to build and how we want to get there. We should aim for the transition to a different kind of lifestyle and society to take place in an organized and orderly fashion in which no one need suffer extreme hardship, gross breaches of basic human rights, or starvation.

In this chapter we take our cues for a safe and steady long-term future from Japan's recent Edo Period, 1603–1867, which was a time of relative stability.

Whilst the Edo economy was almost entirely self-sufficient, today Japan imports 60 percent of its food calories. Its population has quadrupled under these conditions which are, of course, associated with plentiful imports of coal, petroleum, natural gas, and uranium.

Historical statistics tell us that before imports of food began, traditional agriculture probably supported a fairly stable population of about 30–33 million on the Japanese archipelago from around 1720[3] to the Meiji Restoration in 1868, when the Edo Period ends.[4] At that time there were roughly eight people per hectare (ha), or an average of 0.125 ha of arable land per person in a clan-based peasant economy with some feudal features stemming from a military class but without widespread serfdom or slavery.[5]

In Edo Japan, an average hectare inhabited by a family of six to ten members would contain all or most elements for self-sufficiency.

Some examples of land-use arrangements at this population density have been preserved in old residences of a noble class. A "*hanshu*" or clan-lord granted such residences, called "*yashiki*" to the samurai warriors, or "*bushi*," in his retinue who were then expected to be self-supporting.

One such *yashiki*, a 6,000 m^2 *igune*, or land with a forested area, dating from 1774, can be seen in what is now Wakabayashi Ku in Sendai City (northeast Japan). Vegetable gardens occupy about half the total area, a house and other buildings take up another quarter, and the remaining quarter is covered with trees.[6] Rice, grown on fields nearby, made each *yashiki* part of a self-sufficient community organized along clan-based lines.

In the eighteenth century, when rice yields were around 1.5 tonne/ha, a family of ten required about 0.8–1.0 ha of paddy field for rice. An adult consumed an average of one *koku* of rice (roughly equivalent to 140 kg) annually, so one *chôbu* of paddy land (almost the equivalent of a hectare) would support ten adults. Local lords knew the potential population (and fighting force) of their territory simply by knowing the area of their paddy fields. Political and military power was measured in *koku* during the Edo Period.

Surrounding the house in Sendai there are 169 trees. Over half of them are kinds of cedar, cypress, persimmon, plum, and pine, with additional yew, paulownia, chestnut, willow, citron, cherry, walnut, fig, and others. Unusually for Japan, there is no bamboo, which is fast-growing and provides an extremely versatile material, but most of the uses which you could think of for trees are represented; food, house-building, furniture-making, tool-making, fuel for heating and cooking. Tree leaves provide compost and trees act as wind and snow breaks. They also enhance the water-holding potential of the ground.[7]

Far larger *yashiki* were created in the Santome district of what is now Tokorozawa City and Miyoshi Town in Saitama Prefecture (just north of Tokyo) in 1696, some of which are still in existence today.[8] Each consists of about 4.85 ha in a block of land 72 m by 675 m. A road runs along one end of the block with the house, surrounded by trees, on about 0.6 ha. In the center of the block there are about 2.7 ha of upland fields. At the end away from the road and house is a wooded area (*heichirin*) of about 1.5 ha. This *yashiki* was designed with the difference that its samurai farmers (*gôshi*) were expected to produce surpluses which could be traded for rice. Famed for its sweet potatoes (*satsuma imo*), the area was an important food producer in the Edo Period and during the food shortage following the end of the war in 1945.

Japanese culture has thus preserved in the *yashiki* the memory of what basic self-sufficiency is, what it looks like and how it is done, for Japanese people today, and *yashikis* are still written about in newspapers, magazines, and books.

Few Japanese, however, would conceive that a "return" to a similar form of lifestyle might be necessary within the lifetimes of people alive today – by around 2050.

For a solution to be based on Edo-type land-use structure and social division of work, however, the current population to arable land ratio, 29 persons per hectare (cap/ha), is far too high. The numbers would have to come down to around the 8–10 cap/ha (0.1–0.125 ha/cap) hypothesized above. Or perhaps the internationally sourced approximate 0.134 ha/cap (7.5 cap/ha) which now provides the Japanese population with food[9] would be a realistic ideal to work towards locally.

The problem of feeding such a large population on so little land would be assisted by improvements in alternative agriculture, which have kept pace and possibly now exceed those of industrialized agriculture. At today's yields, the rice for a ten-member family would be grown on 0.25 ha (at a yield of 5 tonne/ha and an average consumption of 120 kg/cap/yr of rice), in contrast to eighteenth-century rice yields of the Edo Period, which were only about 1.5 tonne/ha.

Major factors needing to be adjusted to each other in order to tailor an Edo-type land-use solution would be population numbers, land-use availability, social division of work, consumption, and technology.

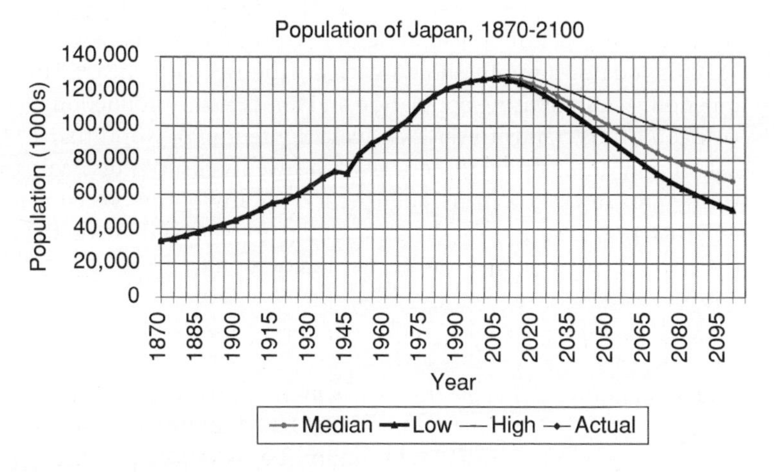

Figure 22.1 Population of Japan projections

Source: National Institute of Population and Social Security Research, www.ipss.go.jp/.

Japan's population has roughly quadrupled (127,780,000 ÷ 30,000,000) in the last 135 years (see Figure 22.1).

In 2004 it peaked at 127.78 million.[10] To discuss its future food security, we need to know roughly what Japan's population is likely to be over the next 50–100 years.

The IPSSR population projections for Japan's population to 2100 imply a return to the early Showa (1925) population level of around 60 million between the 2080s and the early decades of the twenty-second century, if trends persist.

FUTURE WORLD POPULATION AND FOOD PRODUCTION

The lowest projection of the 2006 UN world population revision shows a peak world population of 7.87 billion around 2040 (now about 6.6 billion).[11] Further severe strains on cropland and grain production in the mid-term future seem inevitable.

World grain production per capita has stagnated at around 300 kg/cap over the last two decades and shows signs of further decline. For the five decades since the middle of the twentieth century, global population has been supported by scientific advances in agriculture, largely reliant on fossil fuels. Yield rises accompanying agricultural modernization with its use of chemical fertilizers, agricultural chemicals, hybrid and high-yielding seed varieties, mechanization, irrigation, large-scale monocropping, and so on, appear to be slowing. GM crops do not currently show any great promise in this area.[12] The impending clash of rising population and limits to the production of basic foodstuffs trends towards a severe tightening of world traded grain supply during the early decades of the twenty-first century. With world grain stocks lower in 2006/07 than for any of the preceding ten years, now is surely not the time for Japan to be banking on the international grain market for its supplies of food.[13]

JAPAN'S AGRICULTURAL LAND RESOURCES[14]

Total farmland area, total planted area, and the cropping rate declined from a high in 1956 of 6.012 million ha, 8.27 million ha and 137.6 percent, to 4.69 million ha, 4.38 million ha, and 93.4 percent in 2005.[15]

Japan lost 43,063 ha of farmland to other uses in 1998. In 48.4 percent, cultivation has been abandoned. 26 percent is now occupied by housing, 3.7 percent by factories, and 6.2 percent by roads and railways. Agricultural and forestry roads occupy a further 1.6 percent and forestry has taken over 2.2 percent. Natural calamities account for 4.9 percent.[16] In the five years 2001 to 2005, 160,500 ha of farmland were taken out of production (21,854 ha were added).[17]

With the current cropping rate at about 93 percent, and a decline in per capita rice consumption in Japan (118 kg/cap in 1962, 61.5 kg/cap in 2004),[18] Japan has a set-aside "adjustment" scheme ("*gentan*") to prevent the creation of a local surplus since Japanese rice is too expensive for the export market. An average of 200,000 ha of paddy land was not planted in the four years 1995–98.[19] That land is taken out of cultivation despite huge agricultural imports (including rice) angers many Japanese people and represents a danger in the light of world food trends.

The Japanese Ministry of Agriculture, Forestry, and Fisheries (JMAFF) has calculated the agricultural land needed to produce the food that Japan consumes to be about 17 million ha.[20] This would agree with a generous Edo model, giving the approximate farmland per person used to grow the total food supplied to 127,780,000 Japanese as 17,000,000 ÷ 127,780,000 = 0.133 ha (or 7.5 cap/ha), which Japan could aim for as future ideal.

What is the reality? Efficient land-use results in a cropping rate higher than 100 percent, which means that total planted areas will usually exceed total farmland. Based on the historic 1956 high cropping rate of 137 percent, and improvements in non-industrialized agriculture, 140 percent is probably the practical maximum for Japan as large parts of the north of the country are snowbound in the winter months. Total farmland area for 1961 was an historic high at 6.086 million ha. A theoretical maximum for Japan's total farmland area and total planted area could be calculated as 6,086,000 ha × 140% = 8,520,400 ha, or about 8.5 million ha. At peak population, the maximum planted area per capita would be approximately 8,500,000 ha ÷ 127,780,000 = 0.0665 ha (or 15 cap/ha), which would mean 50 percent more people per hectare than in our hypothesized Edo Period.

In fact the current situation falls far short of that historic recent potential. As seen above, in 2005 Japanese farmland area approximated 4.69 million ha with a cropping rate around 93.4 percent. The planted area per capita at peak population was therefore 4,690,000 ha × 93.4% ÷ 127,780,000 = 0.0343 ha (or 29 cap/ha), something like three to four times the number of people per hectare of Edo.

SOCIAL DIVISION OF WORK

Itakura estimates that in Edo Period Japan about 85 percent of the population were "*nomin,*" or farming people.[21] In 1903, 64 percent of households farmed; 69.6 percent full time and 30.4 percent of them part time.[22]

In 1920, half the working population and a quarter of the total population were working in agriculture. At the beginning of the era of fast economic growth in Japan (1960), these numbers were about 30 percent and 15 percent. However, by the late 1990s this figure had dropped to about 6 percent and 3 percent respectively.

The 2 percent of Japanese actively involved in farming today are aging rapidly. In 2005 more than 61 percent of Japan's full-time commercial (male) farmers were over 65 and more than 86 percent were over 50. With only 169,000 farmers in this category under the age of 50 in this country of 127 million, Japan's systemic vulnerability to a global food shortage is little short of critical.[23]

FOOD CONSUMPTION

Diet change in Japan in the twentieth century

During the Edo Period, when Japan's population numbered around 33 million, total cereal dependency was probably around 90 percent, with 6 percent provided by soybeans and azuki beans, and a further 6 percent by potatoes and sweet potatoes, plus fresh vegetables, fruit, fish, and meat, when available and/or in season.[24]

Remarkable changes took place in the Japanese diet from the 1930s to the 1990s. The increases and decreases are symbolic of the changes that have taken place in Japan over the last 70 years (the mid-1930s was the period when buses, trains, and telephones were beginning to change patterns of life in rural Japan) and over the last 40 years, since the inception of the drive to industrialization and economic growth.

With the exception of wheat consumption, which rose four-fold, direct cereal consumption decreased generally. Intake halved of the traditional soybean foods, *miso* and soy sauce. Consumption of milk and other dairy produce increased 28-fold (four-fold since 1960), along with oil and fat, which rose 15-fold (over three-fold since 1960); meat, which increased 14-fold (six-fold since 1960); eggs, which rose over seven-fold; and fish, which increased nearly four-fold. More than twice as much fruit was eaten.[25] Since 1990 domestic production of cheese has increased nearly 120 percent,[26] all part of the ongoing change (Westernization) of the Japanese diet that has been taking place over the last half-century.

Today's Westernized Japanese diet represents a move away from cereals into animal protein foods. In fact, the total consumption of cereals has not decreased – the cereals are simply "processed" through livestock to provide food in the form of animal protein.

What food does Japan still produce for itself?

Japan could easily have been self-sufficient in food in 1960, but is now grossly dependent on the international market for food supplies. Since the late 1990s, 60 percent of food calories consumed in Japan are imported.[27] Japan ranks about 130 in the world, far below Pakistan, Nigeria, Indonesia, Bangladesh, and Brazil in grain self-sufficiency.

Production of the food staples – rice and soybeans, as well as other cereals – has predictably diminished along with farmland and levels of consumption. Rice is one crop for which Japan can easily be self-sufficient. At the end of the nineteenth century, Japan planted about 2.6 million ha of wet rice per year, but yields were about 2 tonnes/ha. From 1920 to 1969, planted area hovered around the 3 million ha mark, but yields rose from about 3 tonnes/ha to about 4.5 tonnes/ha. Planted area in 2004 was around 1.7 million ha, yield being around 5.2 tonnes/ha.[28]

Planted area for soybeans was a high of over 400,000 ha in the 1870s and peaked in 1910 at 470,000 ha, but by 1995 it was under 69,000 ha. A complicating factor for adducing food needs is that soybeans are now imported as livestock feed.

Self-sufficiency in soybeans has fallen steadily since 1930 and is now around 20 percent, with annual imports since the early 1980s in the 4.5–5 million tonne region.[29]

In areas of Japan with fairly mild winters, and where the land was not snowbound in the winter months, winter wheat and barley were often grown on paddy land as a winter crops. Total production of wheat and barley, 3 million tonnes in 1913 and 3.8 million tonnes in the late 1950s, has plummeted to around 1 million tonnes or less since 1970.[30]

Japan still produces about 80 percent of vegetables, 39 percent of the fruit,[31] 44 percent of beef, 51 percent of pork,[32] and 67 percent of the dairy products[33] it consumes.[34]

Self-sufficient in fish (though at lower levels of per capita consumption) up to around 1980, since then Japan has had to maintain, and even raise, per capita fish consumption by subsidizing its ever-declining fish catches with imports.

Japan has gone from potential or near self-sufficiency in livestock feed in 1960 to about 50 percent overseas dependency in just under 40 years.

Outlook for Japan's local food production

Fresh vegetable imports in recent years are driving down prices, seriously affecting production, farmers' livelihoods and self-sufficiency in vegetables. About 60 percent of imports are from China.[35]

Between them Japan and China are taking huge amounts of fish from the seas, just over 30 percent of the world fish catch in

1995. Note that sea fishing is essentially all fossil energy: ships, fuel, machinery, and equipment. Fish (shrimp, and so on) farming is similar, with an emphasis on feed, chemicals, and antibiotics. Japan does not just eat the fish it takes, but, as a major importer of fish, it *consumes* 30 percent of the world's ocean fish catch.[36]

If fewer animal products were consumed, less feed would be necessary. Further, if land used to produce feed crops were used for food crops, self-sufficiency in human food would rise. It all depends on the diet choices that people make, but in a food crisis they will always choose to consume crops directly rather than to pass them through livestock to "process" them into animal products. Animal protein can still be consumed, but it would have to come from the raising of goats, chickens, ducks, rabbits, and other livestock that does not compete with humans for food.

The Japanese can probably get by without eating imported fruit. Ceasing meat imports should not have a serious effect, and reduction in production of meat could help availability of other foodstuffs.

Japan's high population density and large areas of mountainous forested land place extreme pressure on those relatively flat, easily habitable and cultivated areas. In a food and energy crisis, clearing of flat and lower forested areas is likely to occur.

Political factors

Declines in primary production in Japan since the 1960s coincide with the era of fast economic growth. Japanese agriculture was dismantled in the 1960s as the country transformed itself into one of the world's leading industrial powers. The main feature of this era was modernization of agriculture in order to raise labor productivity to release the labor force from the land for work in manufacturing. In addition, the US targeted Japan as a market for its surplus grains.

As the population moved towards the cities and away from agriculture, and as cheap imported grain came onto the market from the US, it became both uneconomical and impractical to continue the traditional practice of growing grain in winter crops on the rice-paddy land. This was also expedient politically as it was necessary to import goods from overseas to balance the burgeoning exports.

Globalization of the market, massive population growth and expansion of infrastructure, affected rice production in Japan. Whilst rice imports are mandatory under the World Trade Organization, the extremely high cost of land in Japan has driven up all other costs, making production costs of rice in Japan much higher than for internationally traded rice. Price support for free (voluntarily marketed) market rice was abolished in 1995, and the average wholesale price of rice in 2005 was 25 percent lower than ten years before.[37] This has seriously affected the livelihood of farmers and their will to produce Japan's national staple food.

Japan is about as reliant on the US for grain commodities such as wheat, kaoliang, maize, and soybeans as it is on the Middle East for oil and natural gas, importing roughly 60 million tonnes of food-related items per year, 27 million tonnes as grain, including livestock feed.[38]

On-farm energy use

A study of energy inputs to wet rice farming in Japan by Taketoshi Udagawa in 1976 shows that while energy output (the food energy value of the rice) exceeded energy inputs up to the late 1950s, further additions of energy subsidies (for example, machinery, chemical fertilizers, and so on) enabled decreased inputs of labor and animal power. The labor power was then freed to transfer from agriculture to manufacturing industry or to the service sector, thereby allowing the Japanese "economic miracle" to take place. At the same time rice yields improved, but the price was five times the amount of energy input for a 150 percent increase in yield and the benefit of labor input reduction. The result was that nearly 200 gigajoules (GJ) were invested annually in every hectare of rice cultivated.

Energy input for machinery and chemical fertilizers per ha in Japan is an amazing 30 times greater than in the US, perhaps because of the very small size – less than 1 ha – of most Japanese farms, but each with its own tractor and rice planter. Pimentel shows an average of 2.93 GJ/ha (700,000 kilocalories per hectare (kcal/ha)) for machinery in the US in 1977, where Udagawa's figure for Japan in 1975 is a staggering 88.8 GJ/ha (21,201,000 kcal/ha)! In a similar calculation for fertilizers the average for the US in 1977 was 8.8 GJ/ha, compared to 27.2 GJ/ha given by Udagawa for Japan in 1975.[39]

Nitrogen pollution due to overuse of fertilizers

Up to two-thirds of chemical fertilizer leaches into groundwater, lakes, and rivers. The associated nitrogen pollution feeds algal blooms which subsequently die and lead to de-oxygenation of surface waters.

The nitrogen (N) content of imported food for humans and animals in Japan has been estimated at 900,000 tonnes, with a further 700,000 tonnes produced domestically, giving a total of 1.6 million tonnes of N.[40] Nearly 850,000 tonnes of chemically active N in nitrogenous chemical fertilizers were used in Japan in 1996.[41] In addition, Japanese farm animals excreted a total of 72 million tonnes of liquid and solid waste in 1998.[42] Human waste from the giant Japanese conurbations of Saitama-Tokyo-Yokohama-Kanagawa, Osaka-Kyoto-Kobe, and so on, is simply flushed out into the oceans, evoking the image of the water and soil fertility of the American breadbasket being flushed down the toilets of the capital and out into Tokyo Bay.

Since Japan is in huge nitrogen excess, it should not have to use chemical fertilizers at all. Working towards that goal will help prevent nitrogen pollution, provide healthier food, decrease fossil subsidies to farming, drawdown on imported energy, and provide employment in a period of massive economic restructure implied by fundamental fossil energy supply problems.

Off-farm food-related energy use

More than ten units of fossil energy are embodied in each unit of food energy consumed in economically advanced countries today. Of that about 20 percent is used on farms in direct agricultural production of food crops and animal foodstuffs, and about 80 percent off-farm for food processing, transport, retail, preparation, and so on.[43]

Perhaps 15–16 percent of Japan's final energy consumption is used in food-related activities, up to 4 percent being in direct on-farm use.[44] In addition, Japan *pays for* energy consumed in the production and transportation of imported food, in the order of 60 percent of the food consumed in Japan.

To provide the annual average per capita food requirement of about 4 GJ of food energy for the Japanese population, 4GJ × 127.78 million = about 510 petajoules (PJ; 10^{15} joules), would require 510 PJ × 10 = about 5.1 exajoules (EJ; 10^{18} joules) (final energy). Final energy averages about 70 percent efficiency, so this would be the equivalent of approximately 7.3 EJ in primary energy terms. The 2005 primary energy supply of Japan was approximately 22.7 EJ. So the equivalent of about 33 percent of Japan's total primary energy budget was being used to supply food to the Japanese population, with 20 percent of that used directly in agriculture. Approximately half of that was used in-country, since, as we know, the 40 percent of Japanese food produced in Japan is much more energy-intensive than overseas food production.

Japan's energy-import dependency

Whilst the world's dependence on oil for its primary energy consumption is about 39 percent, Japan's was 49.7 percent in 2005 (see Table 22.1).[45] Japan relies on Middle East oil-producing countries for around 89 percent of its oil[46] and was the world's third largest oil consumer in 2005, after the US and China, and the second largest importer.[47] Domestic primary energy production is extremely low, being about 16 percent in 2005 if electricity from nuclear power is counted as domestic production, and about 4.5 percent if it is considered to be reliant on imports of uranium. Thus if the approximately 2.6 EJ of nuclear electricity production being counted as domestic production is counted as an energy import, the import dependency for energy supplies rises to around 95 percent.[48]

Table 22.1 Japan's primary energy consumption 2005

Total	*Oil*	*Coal*	*LNG*	*Nuclear*	*Hydropower & geothermal*	*Renewable*
22.71 EJ	49.7%	20.6%	13.8%	11.6%	3.2%	1.1%

Source: Energy Conservation Center, *EDMC Handbook*, 2007, pp. 19, 26.

Japan imports about 96 percent of natural gas consumed, all in the form of liquefied natural gas (LNG) which is used mostly for electric power generation or as feedstock for petrochemical manufacture.[49]

Japan is by far the world's largest importer of steam coal, mainly for power generation, paper pulp, and cement production. Japan also is the world's largest importer of coking coal for its steel industry. Overall, Japan accounts for over 20 percent of total world coal imports.[50] Recently a world coal peak has been forecast for as early as 2025.[51]

In 2005, 55 nuclear power plants produced about 30 percent of Japan's electricity. Controversy over the siting of new plants, prevention of nuclear accidents, nuclear waste disposal, and the decommissioning of obsolete plants impedes further development of Japan's nuclear power capacity. A recent paper predicts a peak in world uranium production well before 2050.[52] There is also doubt that nuclear power plants, like other kinds of big power plant, can be operated for any length of time without fossil energy supplies.[53]

Japan's capacity for energy self-sufficiency and renewable technologies

It is always possible that some new or old fuel will become plentifully accessible with new technology, such as fusion or ways to exploit the gas hydrates in Japanese waters[54] but such power sources are only in the hypothetical stage and may remain there for the foreseeable future.

Official estimates by the Japanese Resource and Energy Agency are that renewable energy, in the form of solar, wind, or ocean power, provided 1.1 percent of Japan's primary energy in 2003, and will provide 3.1 percent in 2010.[55] Hardly enough to keep the wheels of industry turning.

Let's imagine a Japan which has only very small amounts of usable fossil fuels and therefore extremely low levels of industrial activity compared with now. That also means that the ability to manufacture solar (PV) panels, troughs or water heaters, wind power generators, wave power generators, conventional geothermal power generators, fuel cells, and other such "renewable energy" equipment in any quantity is effectively nil. That could be where Japan will be in 2050, or even sooner, depending on the availability of oil and natural gas in the coming decades.

It is very important to use current fossil reserves wisely along with the "intermediate" renewables (solar, wind, oceanic, geothermal) to help make a slow transition to the "new society" bearable. So, what long-term renewable resources does Japan possess?

Where available, geothermal energy could be very useful in agriculture as heat for winter cultivation or as electricity for agricultural machinery, such as dryers and threshers. Japan possesses 18 geothermal generating plants with a total of 530 megawatts (MW). Most are in northeast Japan or in Kyushu, with one plant on Hachijojima, an island south of Tokyo in the Izu Islands. Two more with a total power of 23 MW are planned.[56]

Despite some contemporary small-scale experiments with biodiesel and other biofuels,[57] Japan's population to arable land ratio makes biofuel even more of a luxury option than it is in most countries (see Alice Friedemann's chapter, "Peak Soil" in this volume).

Watermills and windmills provided a large part of Japan's primary motive power until well into the nineteenth century and Japan currently generates around 10 percent of its electricity from hydropower – around 100 terawatt (10^{12}) hours (TWh). There are plans to extend hydropower further, but the limit is probably about 140 TWh.[58] Hydropower is an obvious option in high rainfall mountainous areas, such as Japan, which is able to take advantage of hydropower from streams and dams. The larger the dam, however, the greater the likely environmental impact and difficulty and expense to build. Small, localized hydropower can be environmentally friendly, although in any one area, the number of locations that can be exploited for hydropower will be limited. Further, although hydropower is a potential energy source, electricity production requires the manufacture of generators, a distribution system, and end-use machinery. This may prevent major hydropower schemes from being a viable option in the distant future.

Like France, Japan developed in times past its plentiful supply of rivers, streams and canals for transport, irrigation and conservative technology like waterwheels. To fully reutilize this asset would require the dismantling of enormous quantities of concrete infrastructure which have lined these waterways since the mid twentieth century.

Agricultural research using ten-year averages has now shown repeatedly that legume-based systems are as environmentally superior and just as productive as chemical-industrial systems.[59] What matters is the fertility of the soil, something *not* guaranteed by chemical fertilizers.

Japan already has an alternative "integrated duck and rice cultivation system." The ducks ("*aigamo*" in Japanese) are allowed to stay and swim around in the rice field from shortly after seedling transplanting. They eat any insects and the weeds as they sprout but avoid the rice seedlings as they do not like the rice plant's stiff, high-silica leaves. The ducks' excreta then provide nutrients for the rice growth. This system is sometimes used with the very fast-growing, nitrogen-fixing floating grass, azolla, which the ducks will feed on. Sometimes fish, such as loach are added which the ducks do not eat because they cannot see the fish beneath the azolla. A well-known practitioner husband and wife team of this system, Mr. Takao Furuno and his wife in Keisen Town, Fukuoka Prefecture, has obtained yields of 6,470 kg/ha of unpolished rice, where the average rice yield in his area is 3,830 kg/ha.[60] Furuno uses no chemical fertilizers or pesticides. In fact, if you visit him, you have to watch where you walk as his neighbors use chemicals and he does not want any of them trudged near his fields. The key determinants of his high yields are animal excreta and soil fertility, not chemical fertilizers.

Agricultural use accounts for about two-thirds of the utilized water in Japan each year. Being 65 percent forested, mostly in mountainous areas, rain is filtered through forest soils and flows down natural streams and rivers to the sea. Around 13–14 percent of the rain is captured on the way, two-thirds of that being used in agriculture. Much of that flows through paddy fields and associated irrigation systems, which perform a water adjustment function for the whole water system. In a natural (unpolluted) ecosystem, the clear, filtered water brings nutrients (especially minerals) to the sea, and this enhances the growth of living material in the sea. In several locations in Japan, notably in the north, coastal people, who have seen their fishing catches dwindle over the past few decades have heroically begun to revive local fishing grounds through organized tree-plantings on local mountainsides.[61]

Japan has a temperate climate, though with a large north–south variation, with two periods of heavier rainfall each year, in June and September, when a cold front moves north to bring on the summer, or south to usher in the fall. The June rains ("*tsuyu*," or "*baiu*") are important for rice cultivation and now also for the provision of water to the cities as they help to fill the dams.

Twice in the 1990s, irregular rainfall resulted in near-crises for Japan; in 1993 when the rains continued for over a month longer than usual, and 1994, when there was too little rain. The almost non-existent summer in 1993 resulted in the worst rice crop in Japan since 1945. Rice had to be imported on an emergency basis from the US, Thailand, and South Korea, causing prices to rise on the international rice market, and providing a convenient precedent for Japan to be finally forced in the Uruguay Round of the General Agreement on Tariffs and Trade (GATT) negotiations to allow rice imports.[62] The 1994 drought resulted in severe water shortages in some parts of western Japan when dams actually went dry.[63] Although relatively water-rich, Japan's food production is vulnerable to extreme climatic swings. This may be exacerbated by global warming.

Forests

One of Japan's greatest resources is, of course, its forests. Wood was the basis of most of the building and machinery associated with industry and hydrology in Japan's pre-industrial past.

At the end of the 1990s, Japan was the world's largest importer of timber products (approximately 20 percent of the world's total trade volume of timber), with the US not far behind.[64] Ironically, forests cover 24,621,200 ha (246,212 km^2) of Japan's land, which totals 37,652,000 ha (376,520 km^2).[65] Therefore, 65.4 percent, very nearly two-thirds of Japan's land area, is forested. Much forested land is mountainous, therefore less likely to be cleared for other uses.

Koichiro Koike of Shimane University in Matsue City suggests that 60 percent, or 15×10^6 ha, of Japan's forests could be sustainably managed and harvested to provide wood biomass and other products. The energy value of the harvested wood can be calculated as follows:[66]

Average annual growth per ha for Japanese forests	13.8 tonnes
Multiply by 0.4 to eliminate leaves and branches	5.52 tonnes
Multiply by 15×10^6 ha for total harvested weight	82.8×10^6 tonnes
Multiply by 18.84×10^9 J/tonne to obtain energy value in J	1.56 EJ

As Japan's total primary energy supply in 1998 was 22.7 EJ, this 1.56 EJ represents nearly 7 percent of current primary energy – not bad for a potentially fully sustainable resource that is hardly tapped at all now. However, burning is not the only thing you can do with wood. At 127 million people, the 82.8 million tonnes of *trunk wood* harvested would provide 650 kg of wood per person per year, some of which would be needed to construct homes, furniture, and so on. A typical Japanese house is about 100 m^2 in floor area and the wood needed to construct it is about 20 m^3. Since the density of timber grown in Japan is on average 0.6–0.65 g/cm^3, about 12–13 tonnes of wood are needed to construct the house (though there is no particular reason why houses should be this size in the future). If 650 kg of wood were harvested per person per year on average, to build this house would take one person 20 years.[67] That assumes the sole use of timber is house-building, but if there are four people in the family, the house can be rebuilt in 20 years using only a quarter of their "ration" of wood. If houses are built to last for 30 or 40 years, then even at the current population level, housing can be supplied on a sustainable basis from Japan's forests. We should bear in mind though that not all kinds of trees (wood) are suitable for construction; the difference in geographical distribution of forests and time necessary for forest growth and management makes this merely a "theoretical" calculation (see below), and all other requirements for house-building (including carpenters, tools, and so on) may not be freely available.

We still have not taken into account the remaining 60 percent of the growth (8.28 tonnes/ha of leaves and branches) that can be used for fuel or composted for maintaining agricultural land fertility, though some of it should remain in the forest to help maintain soil fertility there.

The problem with all of this biomass is that it will need to travel to where it is to be used, which must mean locally, or within easy reach of a suitable waterway, as other forms of long-distance transportation would use more energy than represented by the wood transported. That means that people who live near forests will be relatively better served. Ibaraki Prefecture, where I live, for example had a population of very nearly 3 million in 1998. Its forested area was 195,200 ha. Performing the above calculation gives 1,078,000 tonnes of sustainably harvestable wood, or 360 kg/cap/year (plus the leaves and branches). People in Hokkaido will be better off (but they will need to burn more as the winters are more severe) while people in Tokyo's Shinjuku District, where there is hardly a tree to be seen, will not be so well off, though it might be easier to find shelter there.

What if there is a severe food shortage? Tokyo Shinbun[68] reports that rural inhabitants of the DPRK (North Korea) are clearing forests to make fields for growing food. The fields thus created become the property of the farmer. A man from North Hamyong Province, in the northeast corner of the DPRK, said in an interview that almost all the forests within 10 km of his village had been cleared and that maize and vegetables were being grown on the new fields. The ownership of the fields was being quietly acknowledged by the authorities and a tax consisting of part of the crops was paid. Forest clearing and the need for fuel had resulted in nearly all the trees being felled, and this was causing floods and mudslides. It is therefore essential that Japanese people learn to appreciate the absolute necessity of maintaining their forests in as large an area and in as good condition as is humanly possible, even if the temptation to cut for fuel and clear for food-producing fields is strong.

Clearcutting followed by the establishment of tree plantations (generally fast-growing conifers) is not a good idea either. Large-scale tree plantations were carried out in many places in Japan in the 1950s and 1960s with deleterious effects.[69] These included the replacement of valuable native ecosystems; destruction of animal and plant biodiversity; depletion of water resources; and alteration of mineral content in water runoff, with a generally adverse effect on local agriculture, reversible only in the timescale of centuries.

What needs to be done (or not done) is reasonably clear. The ability to carry out forest protection and management in Japan in coming decades depends on whether the Japanese can become aware of how much their forests mean to them in terms of long-term survivability. In the case of a sudden food and energy crisis, as occurred in the DPRK in the mid-1990s, forest resources will come under immense pressure. Only a great deal of wisdom and foresight can prevent an ecological tragedy from occurring.

Draft animals

Draft animals for plowing, transport and other agricultural uses will only become generally available when Japan has reinstituted a good supply of work animals and will depend on something like one quarter of arable land being set aside for their needs. Animal sources of motive power to replace fossil energy in Japan would mean water buffalo or horses.

Virtually no work animals exist in the Japanese countryside today. Probably no water buffalo remain anywhere in Japan now, except a few on the Okinawan islands. A 30-year project to provide suitable numbers of work animals to the Japanese countryside should probably have started around 1990, but providing the feed for these animals will compete with land needed to grow food for people. The answer to that problem is that the population must decline to within a suitable number for the area and endowments of the land available to allow for work animals. We cannot possibly provide the current motive power of machinery now provided by fossil resources from the use of human and animal power. When the fossil energy age comes to an end we will be living a different life in a different world. Fossil energy resources have allowed some of us to take it easy for a few years. It will be a little difficult getting back to where the ecosphere was about 200 years ago, but to do this without the help that animals can provide will be very grueling indeed.

Assumptions for a future scenario

In order for Japan to alleviate or avoid severe food crises related to the fossil-energy shortage projected for the early decades of the twenty-first century it would be necessary to reinstate aspects of Japanese agriculture that were lost in the late 1950s and early 1960s.

Based on our assumptions about Japan's future population and the implications of agricultural policies, we can see ways in which local and Japanese government might enhance food and energy self-sufficiency, planning for population, and arable land supply in Japan to the year 2100.

In this scenario we assume that cropping ratio would expand by 1 percent per year up to a maximum sustainable 140 percent; arable land area would be expanded by 20,000 ha per year up to the historical maximum of 6.1 million ha; population would continue to fall according to the current median or low projection scenarios; yields maintained while the transition to low-input, sustainable-recycling, organic farming is carried out in an organized and orderly fashion over the first two to three decades of the twenty-first century; and that forest areas are maintained or expanded slightly, and managed sustainably and conscientiously.

If no unforeseen crises occur during this time (they will, of course), in the low population scenario by 2073 Japan would average 8 cap/ha of arable land in a population of 68 million, and by 2080, it would average 7.5 cap/ha of arable land with a population of 64 million in total. At the median population scenario, the level of 8 cap/ha or 0.125 ha arable land per capita would occur around the year 2100.[70]

A century in the affairs of over 100 million people will never be that simple, but the hypothetical exercise does give a fairly realistic picture of the scale of the task ahead.

It will take around 50 years for the population to fall and for arable land areas to rise to their 1950 levels, at around 11 people per arable ha (0.09 ha/cap). This population to arable land ratio would be survivable, if not comfortable. It would not, however, provide land to feed and graze stock.

The big question is, assuming (as perhaps we should not) that we have until about 2025 to go before a serious crisis occurs (no more shipments of food and energy to Japan), how on earth will Japan make it through the following 25 years to 2050?

Arable land area per person in Japan at peak population was about 0.0343 ha (29 cap/ha), which would be disastrous if food imports ceased. Japan now urgently needs to enhance population reduction; reduce reliance on fossil energy resources, primarily by reduction of infrastructure expansion and non-essential personal consumption; increase arable land area and planted area; ensure sustainable management of forests; and implement food production techniques that do not rely on imported energy – fossil energy resources.

NOTES

1. This chapter is an update of a very much longer article, "Food and Energy in Japan – How Will Japan Feed itself in the 21st Century?" (2000), which

can be found at www9.ocn.ne.jp/~aslan/fande21e.htm. See that article for more references, a bibliography, more complex data, and detailed discussion of the issues involved.

2. Eisuke Ishikawa, *2050 is the Edo Period*, Kodansha, 1998 (in Japanese; henceforth denoted by "[J]").
3. Japan Reference, JREF, "Edo Period," www.jref.com/culture/edo_period_era.shtml.
4. Kiyonobu Itakura, *How to Look at and Think About History*, Kasetsusha, 1986, p. 138 [J].
5. V. Smil, *General Energetics*, John Wiley and Sons, New York, 1991, p. 240.
6. Tomio Yuuki, *"A Garden for Living" Makes Scenery*, in *Encouraging Japanese Gardening*, Gendai Nogyo Extra Publication, August 2000, pp. 40–9, [J], and Association for Recording the Nanago of Now and Then, "The Old Village of Nanago; Another face of Sendai," Tas Design Studio, 1993, pp. 96, 100 [J].
7. Association for Recording the Nanago of Now and Then, p. 97 [J].
8. Tokyo Shinbun, March 29, 1999, p. 7 [J].
9. "Basic Direction of Agricultural Administration in the 21st Century," *Japan Agricultural Newspaper*, 1998, p. 58.
10. Japan's population figure of 127.77 million in the 2005 Population Census was below the 2004 population estimate (127.78 million). This marked the beginning of a population decline in Japan. IPSSR, www.ipss.go.jp/, see especially www.ipss.go.jp/pp-newest/j/newest03/newest03.pdf [J]; and English information at www.ipss.go.jp/index-e.html, accessed June 2, 2007.
11. United Nations World Population Estimates and Projections, 2006 Revision, http://esa.un.org/unpp/, accessed June 2, 2007.
12. J. Fernandez-Cornejo and Margriet Caswell, *Genetically Engineered Crops in the United States*, USDA/ERS Economic Information Bulletin No. 11, April 2006, www.ers.usda.gov/publications/eib11/eib11.pdf, p. 9, accessed June 2, 2007.
13. NFU news release, "Lowest Food Supplies In 50 or 100 Years: Global Food Crisis Emerging," http://tinyurl.com/339dw6, accessed June 2, 2007.
14. Japanese Ministry of Agriculture, Forestry, and Fisheries (JMAFF), Agriculture White Paper Statistical Appendix, 2006, p. 33. *Complete Showa Statistics*, Vol. 1, Toyo Keizai Shinbun Sha, pp. 164, 169.
15. The relationship between these three numbers is: Total Farmland Area × Cropping Rate = Total Planted Area; (Total Planted Area ÷ Total Farmland Area) × 100 = Cropping Rate. Statistics from JMAFF, Agriculture White Paper Statistical Appendix, 2006, p. 33 [J].
16. JMAFF, Agriculture White Paper Statistical Appendix, 1999, p. 58 [J].
17. JMAFF, Agriculture White Paper Statistical Appendix, 2006, p. 32 [J].
18. JMAFF, Food Balance Charts, 1996, pp. 64–8; JMAFF, Agriculture White Paper Statistical Appendix, 2006, p. 84 [J].
19. JMAFF, Agriculture White Paper, 1999, p. 21 [J].
20. "Basic Direction of Agricultural Administration in the 21st Century," *Japan Agricultural Newspaper*, 1998, p. 58.

21. Kiyonobu Itakura, *Introduction to Japanese History*, Kasetsusha, 1981, p. 13 [J]; Itakura, *How to Look at and Think About History*, p. 138 [J].
22. *Census Tables of the Meiji and Taisho Eras*, Toyo Keizaishinbunsha, Vol. 1, p. 507 [J].
23. JMAFF, Agriculture White Paper Statistical Appendix, 2006, p. 38 [J].
24. Itakura, *How to Look at and Think About History*, p. 78.
25. JMAFF, Food Balance Charts, 1996, p. 64.
26. National Agricultural Research Center for Hokkaido Region, http://cryo.naro.affrc.go.jp/cheese/cheese/index3.html, accessed June 2, 2007.
27. JMAFF, Agriculture White Paper Statistical Appendix, 2006, p. 14 [J].
28. JMAFF, statistics, http://tinyurl.com/yuslcy, accessed June 2, 2007.
29. JMAFF, Crop statistics, 1996, p. 162; JMAFF, Food Balance Charts, 1996, pp. 120, 236 [J].
30. JMAFF, Crop Statistics, 1996, pp. 150–7.
31. Ibid., p. 55.
32. Ibid., pp. 56, 89.
33. Ibid., p. 89.
34. JMAFF, Agriculture White Paper Statistical Appendix 2006, p. 89 [J].
35. JMAFF, Crop Statistics, 1996, p. 54.
36. JMAFF, *Toward the Establishment of Trade Rules for the 21st Century that Contribute to the Era of "Diversity and Coexistence,"* www.maff.go.jp/wto/emic2.html, accessed June 1, 2007.
37. Nikkei Shinbun, March 16, 2006, www.nikkei.co.jp/news/kakaku/column/20060316e1j1600a16.html, accessed May 11, 2007. ¥15,236/60 kg in 2005 – nominal yen, but inflation was very low in Japan in this period.
38. JMAFF, Agriculture White Paper Statistical Appendix, 1999, p. 43 [J].
39. Taketoshi Udagawa, "Estimates of Input Energy to Rice Cultivation," *Environment Information Studies*, vol. 5, no. 2 (1976), pp. 73–9 [J]; D. Pimentel, ed., *Handbook of Energy Utilization in Agriculture*, CRC Press, Boca Raton, Florida, 1980, p. 96.
40. IRM Research Association, *What is "IRM,"* 1999, p. 4 [J].
41. JMAFF, Agriculture, Forestry and Fishery Statistics, 1999, p. 175.
42. IRM Research Association, *What is "IRM,"* p. 9 [J].
43. Mario Giampietro and David Pimentel, "The Tightening Conflict: Population, Energy Use, and the Ecology of Agriculture," 1994, www.dieoff.com/page69.htm, accessed June 4, 2007.
44. Based on J. Gever et al., *Beyond Oil*, University Press of Colorado, New York, 1991, who suggested that 10–13 percent of the US energy budget was used in off-farm food-related activities in the early 1980s.
45. The Energy Conservation Center, Japan, *EDMC Handbook of Energy and Economic Statistics in Japan*, 2007, p. 26 [J.]
46. Ibid., p. 162 (figure for 2005) (85.2 percent in 1999, 71.5 percent in 1970).
47. The Energy Conservation Center, Japan, *EDMC Handbook of Energy and Economic Statistics in Japan*, 2005, p. 20 [J].
48. The Energy Conservation Center, *EDMC Handbook of Energy and Economic Statistics in Japan*, 2007, p. 16. Japan's reserves and production of uranium are effectively zero: Energy Watch Group, *Uranium Resources*

and Nuclear Energy, Background Paper prepared by the Energy Watch Group, EWG-Series No. 1/2006, December 2006, p. 28, www.lbst.de/publications/studies__e/2006/EWG-paper_1-06_Uranium-Resources-Nuclear-Energy_03DEC2006.pdf, accessed June 4, 2007.
49. Ministry of Foreign Affairs, Economy Division, Department of Energy Security, 2006, www.mofa.go.jp/mofaj/gaiko/energy/pdfs/j_josei.pdf, [J], accessed June 4, 2007.
50. 162.6 million tonnes in 2002: Nissho Iwai Co., Ltd., *Strategy for Stable Supply of Coal*, 2004, www.meti.go.jp/report/downloadfiles/g40324b40j.pdf, [J], accessed June 4, 2007.
51. Energy Watch Group, "Coal: Resources and Future Production," 2007, p. 17, www.energywatchgroup.org/files/Coalreport.pdf, accessed June 4, 2007.
52. Energy Watch Group, *Uranium Resources and Nuclear Energy*, p. 5.
53. A. Boys, "Japan and the End of Cheap Oil," *New Internationalist Japan*, no. 24 (July 2001), www9.ocn.ne.jp/~aslan/nij.pdf, accessed June 4, 2007.
54. Japanese Resource and Energy Agency website, www.enecho.go.jp/lng/pages/page6.html, [J], accessed July 10, 2000.
55. Japanese Resource and Energy Agency website, www.enecho.go.jp/shinene/pages/page1.html, [J], accessed July 10, 2000.
56. See, for example, the Geothermal Research Society of Japan's English website, wwwsoc.nii.ac.jp/grsj/geothermalinJ/index.html, accessed June 4, 2007.
57. For example, in Aito Town in Shiga Prefecture.
58. Japanese Resource and Energy Agency website, www.enecho.go.jp/suiryoku/pages/page1.html, [J], accessed July 10, 2000.
59. L.E. Drinkwater, P. Wagoner, and M. Sarrantonio, "Legume-Based Cropping Systems Have Reduced Carbon and Nitrogen Losses," *Nature*, vol. 396 (November 19, 1998), pp. 262–5.
60. Mae-Wan Ho, "One Bird – Ten Thousand Treasures," *Ecologist*, vol. 29, no. 6 (October 1999), pp. 339–40; T. Furuno, *The Power of Duck*, Tagari, 2001.
61. For example, Shigeatsu Hatakeyama, *The Forest is the Lover of the Sea*, Hokuto Shuppan, 1994 [J], and *The Forest Protects the Abundance of the Sea*, Tokyo Shinbun, January 25, 1999, p. 5 [J].
62. Takashi Doi, *The Great Famine Era*, Yomiuri Shinbun Sha, 1996, pp. 29–32 [J].
63. A. Boys, "Drought Hits Japan," *Green International*, issue 66 (December 1994).
64. JMAFF, Forestry White Paper, 1998, pp. 202, 203 [J]; JMAFF, Agriculture, Forestry and Fishery Statistics, 1999, pp. 371, 372 [J].
65. JMAFF, Agriculture, Forestry and Fishery Statistics, 1999, pp. 1, 326 [J]. Area refers to land area only; forested area is for 1990.
66. Interview with Koichiro Koike at Shimane University, Matsue City, February 7, 2000; Sadakichi Kishimoto, "Thinking About the Energy of Forests," *Sohbun*, 1981, pp. 31, 32, 59 [J].
67. Kishimoto, "Thinking About the Energy of Forests," p. 49. The 20 m^3 figure is from Mr Tomoji Hirata, architect, of Hitachiomiya City, Ibaraki Prefecture, Japan.

68. Tokyo Shinbun, August 13, 2000.
69. World Rainforest Movement (WRM), WRM Briefings, "Tree Plantations as Sinks Must be Sunk", www.wrm.org.uy/publications/briefings/CCC.html#sinks, accessed June 4, 2007; interview with Sato Kisaku, Nikaho Town, Akita Prefecture, September 3, 1999.
70. A. Boys, "Food and Energy in Japan – How Will Japan Feed Itself in the 21st Century?" 2000, p. 58, www9.ocn.ne.jp/~aslan/fande21e.htm.

23
The Simpler Way

Ted Trainer

Before discussing energy arrangements compatible with an ecologically sustainable society, it is necessary to clarify the extent to which rich countries presently exceed the levels of consumption that could be maintained indefinitely, or extended to all people. The overshoot is enormous and, accordingly, the amount of energy use in a sustainable society will have to be a small fraction of the amount we take for granted in consumer society today. It follows that a sustainable society cannot be achieved without very radical changes in lifestyles, systems of land use, patterns of settlement, the economy, and social values.

The following is only the briefest review of major arguments supporting the "limits to growth" outlook on our global situation:[1]

- If energy production were increased to the point where a world population of 9 billion people consumed energy at current per capita rates of the rich world, *all* estimated fossil fuel reserves (*including* an assumed 2,000 billion tonnes of coal) would be totally exhausted within about 40 years.
- The "footprint of productive land" required by each person to sustain a rich-world lifestyle is around 7–12 hectares. The per capita amount of productive land on the planet is only about 1.2 hectares, and by 2050 will probably be close to 0.8 hectares.
- Climate scientists inform us that if the carbon dioxide content of the atmosphere is to be below *twice* the pre-industrial level (many scientists argue it should be far lower than that), total emissions must be held below 9 billion tons per year. For 9 billion people, that is one ton per person. Present US and Australian per capita emission rates are around 16 tonnes per year, and for Australia, if land clearing is included, the figure is 27 tonnes!

These kinds of consideration provide impressive support for the conclusion that energy consumption and resource demand patterns in rich countries are already far beyond sustainable limits. Yet virtually all countries seek economic growth, and ignore any question of limits.

I have argued in a number of works that renewable energy sources and energy conservation cannot substitute for fossil-based energy supplies, and that technical advances, a "factor four" transition and "dematerialization" are not capable of solving the problem.[2]

THE "SIMPLER WAY" ALTERNATIVE

Given this "limits to growth" context, it should not be surprising that the discussion of desirable social forms leads to extremely radical conclusions. It is clear not only that a sustainable society cannot have a growth economy, but that consumption standards must become far lower than they are in rich countries at present. Nevertheless, advocates of the "Simpler Way" firmly believe that it could provide all with a higher *quality* of life than is typical of rich countries today.

The following is a broad outline of this vision. The intention is not to detail energy sources, production quantities, and technologies utilized, but to sketch the kind of society we must shift towards if we are to solve global problems. Of course, this kind of alternative society would enable energy demand to be cut to far below present levels.

If the limits are as savage as they increasingly seem, then the essential and inescapable principles for a sustainable, alternate society must include the following six points:

1. A simpler, non-affluent way of life

We must aim at producing and consuming only as much as we need for comfortable and convenient living standards. We must phase out many entire industries. But living materially simply does not mean deprivation or hardship. There is no need to cut back on production of anything we need for a very comfortable, convenient, and enjoyable way of life. The goal should be to be satisfied with what is sufficient.

2. The development of many small-scale, highly self-sufficient local economies

Most basic necessities should be produced very close to where we live. Declining energy supplies will prevent present levels of transport and packaging from being maintained, making economic decentralization a key requirement. We need to convert our neighborhoods, suburbs, and towns into small, thriving local economies which produce most of the goods and services they need, using local resources wherever possible.

Every suburb would have many small productive enterprises such as farms, dairies, local bakeries, and potteries. Many existing economic entities would remain, but their operations would be decentralized as much as possible, with workers living close to their place of work, enabling most of us to get to work by bicycle or on foot. Many farms could be backyard and hobby businesses. A high proportion of our honey, eggs, crockery, vegetables, furniture, fruit, clothing, fish, and poultry could come from very small, local family businesses and cooperatives. We would, however, retain some mass production facilities, but many items of general necessity such as furniture and crockery could in the main be produced through craft-working. It is far more satisfying to produce things using craft-working than in factories.

Market gardens could be located throughout suburbs and even cities – for example, on derelict factory sites and beside railway lines. Having food produced close to where people live would enable nutrients to be recycled back to the soil, through garbage (biogas) gas production units. This is essential for a sustainable society. Two of the most unsustainable aspects of our present agriculture are its heavy dependence on energy inputs and the fact that it takes nutrients from the soil and does not return them.

We should convert one house on each block into a neighborhood workshop. It would include a recycling store, meeting place, leisure resources, craft rooms, barter exchange, and library. Because we will not need the car very much when we reduce and decentralize production, we could dig up many roads, thereby making perhaps one-third of a city's area available as communal property. We can plant community orchards and forests and put in community ponds for ducks and fish. Most of your neighborhood could become a permaculture jungle, an "edible landscape" crammed with long-lived, largely self-maintaining productive plants such as fruit and nut trees.

There would also be many varieties of animals living in our suburbs, including an integrated fish-farming industry. Communal woodlots, fruit trees, bamboo clumps, ponds, and meadows would provide many community goods. Local supplies of clay could meet all crockery needs. Similarly, cabinet-making wood might come from local forestry, via one small neighborhood saw bench located in what used to be a car garage.

There is a surprising amount of land in cities that could be used to produce food and other materials. Firstly, there will be home gardening, the most efficient and productive way to provide food. Many flat rooftops can be gardened. Enabling the majority of persons to move from cities to country towns would make for more garden space in cities.

There is immense and largely untapped scope for deriving many materials from plants and other sources that exist or could be developed where we live: bark for tanning, dyes from plants, tar and resins from distilled flue gases, wool, wax, leather, feathers, paint from oil seeds like sunflowers, and many medicines from herbs. Small animals are easily kept within urban neighborhoods, and can yield many products including leather and fertilizer. Much of their feed could consist of recycled kitchen and garden waste. Timber would come from the woodlots and clay from the local pits. Many of these things would come from the commons we should develop in and around our settlements, including orchards, ponds, forests, fields, quarries, bamboo clumps, herb patches, and so on, which would be owned, operated, and maintained by the community.

We could build most of our new housing ourselves, using earth and recycled materials, at a tiny fraction of present housing construction energy cost (and present cash cost).

We would also have decentralized, small-town banks run by elected boards, making our savings available for lending only to socially useful projects in our town or suburb. Local "business incubators" would help small firms to start up with low- or zero-interest loans where appropriate. We would then be in a position to create a dense network of many small firms that would enable unemployed people to start producing to meet those needs that are presently ignored. Because all our local small industries would be owned by people who live in our area, profits would not be siphoned out to distant shareholders but would be spent or reinvested in our area.

This would be a leisure-rich environment. Most suburbs at present are leisure deserts. The alternative neighborhood would be full of interesting things to do, common projects, animal husbandry, small firm activities, gardening, urban forestry and community workshops. Consequently, people would be less inclined to go away on weekends and holidays, again reducing national energy consumption.

Most of the things we need for everyday life in a sustainable society could be produced within a few kilometers of where we live; indeed, most needs would be satisfied by neighborhood production.

Some items, such as radios and stoves, could be produced in factories within ten to twenty kilometers. Perhaps a small city might need one refrigerator manufacturer and repair center. Only a few specialty items might have to be transported hundreds of kilometers from large factories and very few would have to be imported from other countries – for example, high-tech medical equipment. Rational social and economic decisions would have to be made on the location of production facilities for goods that would be exported out of the local region, so that all towns and suburbs can earn a sufficient, small amount of export income to pay for their small import needs.

3. More communal, cooperative, and participatory practices

The third necessary characteristic of a sustainable society is that it must be much more communal, cooperative, and participatory than the society we know today. We must share more things. For example, we could have one stepladder in the neighborhood workshop, rather than one in most or many houses. We would give away surpluses. We would have voluntary community "working bees" to provide most child-minding, nursing, basic education, and care of aged and handicapped people, as well as performing most of the functions that town and local councils presently carry out on our behalf, such as maintaining parks and streets.

The working bees and neighborhood committees would also maintain the many local commons, such as the orchards, woodlots, ponds, clay pits, workshops, windmills, and other local renewable energy supply systems.

There would be a far greater sense of community than there is now. People would know each other and would constantly interact in community projects. One would certainly predict a huge decline in the incidence of loneliness, depression, and similar social problems, and therefore in the cost of providing for people who have turned to drugs or crime, or who suffer stress, anxiety, and depressive illness. It would be a much healthier and happier place to live, especially for young and old people. Markedly reduced rates of anomie, social stress, and suicide can easily be predicted.

There would necessarily be a transition to a radically different form of government: to small-scale, local, and participatory democracy. Most of our local policies and programs could be worked out by elected, unpaid committees, and we could all vote at general assemblies or town meetings on the important decisions concerning our small area. There could still be functions for state and national governments, but relatively few.

4. Alternative technologies

In some areas people could still use as much modern technology as they wished – in medicine and dentistry, metalworking, information technology, and so on – and much research could go into developing better technologies. However, in most areas use would be made of relatively simple, traditional and alternative technologies, because these have far lower resource and ecological impacts, and because they are more enjoyable and convivial. For example, most food will be produced by hand tools from home gardens, small local market gardens, and permacultured "edible landscape" commons. These are the most enjoyable ways to produce the best food. Some farms will use some machinery, but on a small scale.

Water will mostly come from rooftops and pollution-free creeks and landscapes. Much manufacturing will be through crafts, hand-tools and small family firms and cooperatives. Many "services," such as the care of older people, will mostly be given informally and spontaneously within supportive communities, not via bureaucracies and professionals. We would research plant- and earth-based substitutes for some scarce minerals and chemicals. Many more tasks will be performed by human labor, as distinct from machines, such as cutting firewood and producing food, because this is more satisfying and because there will not be much energy available for running machines.

Although the Simpler Way seeks the simplest ways of doing things, it is not ideologically opposed to modern technology. Photovoltaic cells, for instance, are desirable, although they are technically complex. However, the Simpler Way notes that sophisticated modern technology is mostly unnecessary, and that technical progress is of little significance in solving the world's real problems or in providing a high quality of life to all. The key to these objectives is applying *simple* methods to meeting human needs while satisfying ecological requirements, which is not done in the present economy, or in our competitive, individualistic culture.

5. An almost totally new economic system

There is no chance whatsoever of making these changes while we retain the present consumer-capitalist economic system. This is the crucial implication from the "limits to growth" literature. The major global problems we face are primarily due to this economic system. The new economy must be organized to meet the needs of people, the environment, and social cohesion, with a minimum of resource use and energy consumption for a maximum quality of life. This is totally different from an economy driven by profit, market forces, and growth.

The need for small, highly self-sufficient local economies, and for zero economic growth, has been noted. There will be relatively few big firms, little international trade, not much transporting of goods between regions, and very little, if any, role for transnational corporations and banks.

Market forces, free enterprise, and the profit motive might be given a place in an acceptable alternative economy, but they could not be allowed to continue as major determinants of economic affairs. Basic economic priorities and structures must be planned for, provided, and regulated according to what is socially desirable (democratically planned, mostly at the local level; not dictated by huge and distant bureaucracies). However, much of the economy might remain as a carefully regulated and monitored form of "private enterprise" carried on by small firms, households and cooperatives, so long as their goals were not profit-maximization and growth. There would have to be extensive discussion and referenda in deciding how to sympathetically phase out the many unnecessary and wasteful industries that now exist, and how to reclassify and redeploy their workers. Social machinery, especially the economy, is very complicated and problems can easily arise. A great deal of effort will have to go into operating, monitoring, debating, and revising this machinery.

There would be a large and important non-monetary sector of the economy, including giving, mutual aid, volunteer work on committees, working bees, and the supply of free goods from local commons. Working bees' activity could be an effective way to pay "tax" – that is, to contribute to the maintenance of public facilities.

The new economy would probably have a relatively small cash sector, and would allow carefully regulated market forces to operate within it. Most of the important large enterprises, such as railways and steel, would probably be planned and run by collective or public agencies or firms, and at the local level would be operated by community cooperatives. Possibly the largest sector of the new economy would be run by community service cooperatives – the local energy supply or water supply cooperative "firms," for example.

Most of us would live well, with greatly reduced needs for cash income, because we would not need to buy very much. Consequently, many of us might work only one day a week for money, and spend the rest of the week work-playing around our neighborhoods in a wide variety of interesting and useful activities. There would be no unemployment and no poverty, as expressed in the ideal of Israeli Kibbutz settlements. We would have local work-coordination committees, which would make sure that all who wanted work had a share of the work that needed doing in the area. All people could make important economic contributions, even though some might have few educational qualifications or be mentally or physically handicapped, because there would be many simple but crucial jobs to be done in the gardens, workshops, forests, and animal pens. All people could be fully active and valued participants in the economy.

There would be far less need for capital; capital-intensive factories and infrastructures such as roads, dams and power stations would be reduced, because the volume of production and transportation, and quantity of energy needed, would be far less than they are now. There would be fewer types of products. For example, we might decide to have only a few types of radios, televisions, buses and cars, designed to last and to be repaired easily.

Few big firms and little heavy industry would be needed, because there would be much less production, especially of complex and sophisticated goods. Most items would be produced by small family firms and cooperatives in which people would invest their own savings, deriving modest, stable incomes. A few large firms would provide things like steel and railway equipment, and these should be run as public enterprises. Again, their control must be through open and participatory mechanisms, not necessarily by the state. There must be processes whereby all people can constantly monitor and evaluate the performance of public institutions and enterprises.

The focus in the above account has been on the neighborhood and suburban economy. Beyond these local economies there would still be regional, national and international economies, but their activity levels would be far lower than at present.

6. New values

Obviously, the Simpler Way will not be taken unless there is change from the presently dominant values and habits.

There must be a much more collective, less individualistic social philosophy and outlook; a more cooperative and less competitive attitude; a more participatory and socially responsible orientation; and above all, much greater willingness to be satisfied with less, and by what is simple but sufficient.

These are the biggest difficulties facing the transition to a sustainable society. However, it is important to recognize that the society we have now forces us to compete against each other; for example, for jobs, and that people now consume mainly because few other sources of satisfaction or meaning are open to them in consumer-capitalist society.

On the other hand, the Simpler Way offers many satisfactions and rewards: if people can be helped to see this they will be more likely to move away from consumer society. Consider, for example, having far more time outside the economic nexus, and having to work for money only one or two days a week, living in a rich, varied, and supportive community, with interesting, enjoyable, varied, and worthwhile work to do, contributing to the governance of one's community, participating in meetings and decision-making for the good of all. Consider also having much more time for learning and practicing arts and crafts, for personal development, and for community development, participating in many local festivals and celebrations, running a productive, efficient, and highly self-sufficient household and garden. Consider living in a leisure-rich environment, being secure, not having to worry about unemployment or being lonely, not worrying about how you will cope if you are ill or when you are aged. Above all, there is benefit in knowing that you are no longer part of the global problem, because you are living in ways that are sustainable.

Compared with people in the consumer society of today, we would be very poor, wearing old clothes, living in small, sometimes mud-brick houses, and earning very low cash incomes. However, the Simpler Way makes possible a much better quality of life than most people in rich countries experience at present.

There is nothing backward or primitive about the Simpler Way. We would have all the high-tech and modern tools that make sense, such as in medicine, renewable energy technology, public transport and simpler, energy-saving household appliances. We could still have smaller but effective national systems for many things, such as railways and telecommunications. We would also have far *more* resources for science and research, and for education and the arts, than we do now, because we would have liberated those resources presently being wasted, for example in the production of unnecessary items, including arms.

CONCLUSION

The core claim made here is that only if we move, both in rich and poor countries, to something like this vision of the Simpler Way can we expect to achieve a just and sustainable global situation. Only by instituting materially simple, self-sufficient, and cooperative practices within a new, zero-growth economy, can we hope for a high quality of life at much lower levels of energy use.

There is now a global alternative society movement gathering momentum throughout the world, in which many small groups are actually building settlements more or less along the lines outlined above.[3] The fate of the planet depends on how successful this movement will be in creating demonstration settlements, and proving their feasibility before the problems in rich countries become so acute that a reasoned, ordered, and sensible transition becomes impossible.

NOTES

1. For a detailed account see www.arts.unsw.edu.au/tsw/.
2. Ibid.
3. R. Douthwaite, *Short Circuit*, Green Books, Totnes, 1996; W. Schwarz and D. Schwarz, *Living Lightly*, Jon Carpenter, London, 1998; B. Grindheim and D. Kennedy, *Directory of Ecovillages in Europe*, Germany, 1999.

24
In the End: Thermodynamics and the Necessity of Protecting the Natural World

Sheila Newman

Humans already use most of the land on the planet. In many places in the world the competition is between the land-poor and the land-rich. This is a political problem which needs to be solved without further trashing the natural environment. Some systems are more equitable than others and, as discussed in other chapters in this volume ("101 Views from Hubbert's Peak" and "France and Australia After Oil"), the Anglo-Celtic system used in most English-speaking countries is worse than most. We humans have to share the land we already have more equally with each other. If we insist on growing our population then the competition for land will be increasingly severe. We have already taken enough from other creatures and need to give some (a lot) back. Land for wildlife is not a luxury. The perception that it doesn't "do" anything needs scientific countering with a thermodynamic explanation. That explanation is that Life is the only force that can reorder spent energy.

The First Law of Thermodynamics states that energy cannot be created but is never lost. However, the Second Law of Thermodynamics states that energy is transformed by use and that you can never make it how it was before. (You can't have your cake and eat it.) Industrial society provides a good example in biological energy (food) with the idea that a machine can make a sausage out of a pig, but it cannot make a pig out of a sausage. Once you have turned the pig into sausage for human consumption, the energy in the sausage will never be pig again. It will be human waste. The passage of the cake or the pig into something less coherent is part of the usual flow of chemical and physical reactions in a process known as "entropy increase" or a tendency to "disperse."

Otherwise the planet and the atmosphere would be completely filled with sludge and debris. This is the way that ecology and the lifecycles that make up an ecosystem are able to temporarily make order from disorder. Industrial manufacturing can grossly restructure dead things, but life is the only process that is able to do this efficiently, keep the process going, and reproduce itself.

Of course, if you feed a pig a pork sausage, some of that sausage will become pig again. Most people would see, however, that converting a pig into a sausage through an energy-intensive industrial process and then feeding the sausage back to a pig so that the sausage contributed to a miniscule portion of pig flesh is a pretty inefficient way to make pigs. Nonetheless, pork sausages and many other processed foods do find their way back to pigs' troughs. This is quite illustrative of the circular and needlessly wasteful (and cruel) cycles that occur in consumer-industrial societies.

Modern human societies are in fact quite different from those of pre-fossil-fuel human societies and those of other animals.

We modern humans no longer just produce animal waste that is "biodegradable" in a normal ecological cycle. Through extractive technologies we have artificially extended our bodies and amplified our activities, so that we consume quite enormous quantities of material and energy. In the process of digging up the materials and burning the energy to make things with, we also clear almost every other living thing in our paths. The waste carbon, nitrogen, phosphates, sulphur, and other products which our artificial system puts out largely overwhelm the services of the remaining (shrinking) natural ecosystems. Yet the natural ecosystems are the ONLY agents capable of saving us from being buried, suffocated and burned by the physical and chemical interactions of our industrial-society waste.

That is how the second law of thermodynamics can be used to explain why it is vital to allocate increasing space to natural processes. Returning land to wild grass and forests and giving animals their freedom to live naturally is the most positive thing that we humans can do about the accelerating rate of planetary entropy that consumer society multiplied by huge human populations is causing. Entropy comes in the form of increasingly unpredictable climate and in broken, dead, and dying ecosystems.

Large serviceable ecosystems like the Amazon, the great grasslands, coastal waters, coral reefs, indigenous forests, and vast regional chains of animals and plants working in harmony are deteriorating and disappearing because human society and infrastructure are consuming, clearing, fragmenting, and isolating them.

In their place man-made things simplify what existed before. Roads interrupt the living fabric of species interacting. Mines pulverize complex geological features and reduce them to their molecular components. Mines and wells extract, refine, and simplify the geologically processed bodies of ancient plants and animalcules. From these rich sources factories mix the simplified components into soups, pastes, and blocks for building and other materials or as fuel for heating, cooling, machines, and transport. Factory-simplified engineered monocrops, cultivated with one-size-fits-all mass-produced fertilizers destroy living soil and the rich cloak it sustains on the planet's surface. Feedlot farming suppresses individuality by industrialized cruelty in the service of consumerism replacing the awe-inspiring herds and flocks of yore which had their own histories of migrations, navigations, and evolutions. Cities have replaced the mysterious, nurturing and cooling forests, raising local temperatures without making rain. High-rise buildings burn huge amounts of energy and pollute the atmosphere just to keep their temperatures comfortable and their air breathable and to transport people within them by elevators.

These thermodynamic reasons justify the protection of wildlife and natural habitat as necessary for human survival.

People concerned about petroleum decline, pollution, poverty, homelessness, unemployment, and so on, also worry that the survival needs of vast human populations in an era of likely fossil fuel decline will be used to make life even more horrid for other species.

We can see, though, that there is a way that kindness to wildlife and the preservation of habitat is linked to the principles and laws of energy preservation.

As the natural world shrinks it becomes ever more vital to the survival of the human species. Because humans use fossil and other non-biological fuels, overall entropy increases at a much greater rate than it would if we had continued to live without our synthetic infrastructure.

The only thing that can even temporarily recreate some degree of order is life, which creates orderly systems (albeit creatures with finite lifespans but who reproduce) whilst consuming energy. At the moment human beings are increasing entropy a great deal more than the other creatures on the planet, due to the rate at which they draw down upon and burn fossil fuels.

The other creatures in our environment compensate for our activities to some degree. The more of them and the fewer of us, the better for the planet; hence the better for humans who inherit the mess we are making.

Apart from this, the principle of kindness and generosity to our fellow travelers on this planet is a positive one, whereas to reduce them to mere expendable conveniences for our species depraves us.

It's cool to be passionate about nature …

To simply love the source of what we eat, drink and breathe …

Overpopulation and overdevelopment are destroying nature …

Let's talk about it.

Campaign notes for a Sustainable Population
(mostly) by Greg Wood, Rainbow Beach, Australia, 2007

Notes on Contributors

Antony F.F. Boys is British, but for over 30 years has lived in Japan, where he studied the modernization of China's agriculture at Tsukuba University and worked at a number of universities in the Kanto area. He is currently a guest lecturer at Tohoku University Graduate School of Agricultural Science (Sendai) where he teaches one course in food and energy. He has been a regular contributor to Japanese academic journals on the subject of energy and agriculture. He is currently resident in Chiang Mai, Thailand (2007), to research the rotational swidden farming system of the Karen people and to add fluency in the Thai and Karen languages to his already fluent Japanese.

Colin J. Campbell has worked as an oil exploration geologist in many parts of the world, including Australia, Trinidad, Colombia, and New Guinea. In 1968, he joined Amoco in New York to work on global new ventures and resource assessment. In 1984, he became Executive Vice-President of Fina in Norway. Since 1989 he has provided independent advice to governments and major oil companies. He is the author of five books and has published many articles and has lectured widely. Most recently he founded ASPO, an international Association for the Study of Peak Oil.

Dr. Michael Dittmar received his PhD in physics in 1985. He is a researcher at the Swiss Federal Institute of Technology (ETH) Zurich physics department, specializing in particle physics, and works at the European Council for Nuclear Research (CERN). For several years he has given a course "Energy and Environment in the 21st Century" at ETH and is President of the CERN club conCERNed for humanity.

Alice Friedemann has been part of the peak oil community since the first online discussions began on EnergyResources and RunningOnEmpty and has attended many ASPO conferences. She has been interviewed by *Salon* magazine, spoken at UC Berkeley on biofuels, published at EnergyBulletin, Energypulse, The Oil Drum, and other sites too numerous to mention. She has a BS in biology with a chemistry/physics minor from the University of Illinois, Champaign-Urbana, and is a member of the Northern California Science Writers Association. Her website can be found at www.energyskeptic.com.

Mark Jones was an author and novelist with a long-time interest in the political economy of oil. He lived and worked in the Soviet Union and in post-Soviet Russia, working for a time with the *New Statesman*, whilst researching archives newly opened under Gorbachev's policy of Glasnost. He founded the Conference of Socialist Economists and their journal, *Capital and Class*, in the mid-1970s, and in the 1980s he was remanded in custody for publishing Leon Trotsky's writings. He sustained a deep interest and sympathy for communism, and was concerned by the limits to growth and believed that the US economy had an unsustainable but unstoppable need for oil. His novel *Black Lightning*, about the fall of the Soviet Union, was published in 1995 by Victor Gollancz.

Seppo A. Korpela is on the founding team of ASPO-Finland. Born in Finland, he attended primary and secondary school there. In 1972 he received a PhD from the University of Michigan in Ann Arbor. Since then he has been a Professor of Mechanical Engineering at the Ohio State University. His teaching and research interests are in applications of heat transfer. He has contributed to optimizing the gap width of double- and triple-pane windows, and developed a theory for heat transfer by conduction and radiation through fiberglass insulation. He is a member of the Green Building Forum of Columbus, Ohio, a group formed to promote sustainable building practices.

William Ross McCluney is Principal Research Scientist at the Florida Solar Energy Center, a research institute in Cocoa of the University of Central Florida in Orlando, Florida, US. Subsequent to taking his current position in 1976, he taught a graduate course in optical oceanography at Florida Institute of Technology as an adjunct professor and supervised student research work at Florida Tech in oceanography and physics. In 2003–04, Dr. McCluney taught a UCF course on Philosophy, Religion, and the Environment. In 2004 he gave the plenary session lecture to Canada's University of Waterloo Solar Energy Society on justifications for energy efficient and renewable energy technologies. He is the author of *The Environmental Destruction of South Florida* (University of Miami Press, 1971) and the popular textbook, *Introduction to Radiometry and Photometry* (Artech House of Boston, Mass., 1994).

Andrew McKillop conceived the idea for the first edition of *The Final Energy Crisis* (Pluto Press, London, 2005), which he edited (with Sheila Newman). He is a consultant on oil and energy economics, with wide international experience in energy, economic and scientific organizations, in Europe, Asia, the Middle East, and North America. These include the Canada Science Council, the ILO, the European Commission, the European Economic and Social Commission for Asia and South Pacific, and the World Bank. He is a founding member of the Asian chapter of the International Association of Alternative Economics. He has published widely in journals, including *The Ecologist*, the *New Scientist*, and *Le Monde Diplomatique*. In 2007 he was an adviser on Energy Transition funding for Juno Investments, New York.

Sheila Newman is an Australian environmental sociologist who compares social systems. She looks at the connections between different land-tenure systems, population numbers, and energy, comparing population spacing algorithms in other species with inheritance factors in human land-use allocation systems, finding that some systems do more damage than others. She is horrified by the damage that land speculation and politically engineered population growth are inflicting on bushland, wildlife, water catchments, and soil in Australia and elsewhere. As well as writing she uses digital film and animation to communicate how democracy and the environment are being sacrificed to "development" in a mad race against fossil fuel depletion and climate change. Some short films can be accessed at www.youtube.com/QueenieAlexander2000. Recent writing can be accessed at www.candobetter.org/sheila.

Ted Trainer is an academic in the Department of Social Work, University of New South Wales. He has written a number of books on global sustainability and justice issues, including *Abandon Affluence* (Zed Books, London, 1985), *Developed to Death* (Greenprint, London, 1989), *The Nature of Morality* (Avebury, Aldershot, UK, 1991), *The Conserver Society* (Zed Books, London, 1995), *Towards a Sustainable Economy* (Envirobooks, Sydney, 1995), *Saving the Environment: What It Will Take* (University of NSW Press, Sydney, 1998). He is developing an alternative lifestyle education site, Pigface Point, near Sydney. For material that critical global educators might use, see www.arts.unsw.edu.au/tsw/.

Index